SpringerBriefs in Applied Sciences and Technology

Computational Intelligence

Series editor

Janusz Kacprzyk, Systems Research Institute, Polish Academy of Sciences, Warsaw, Poland

The series "Studies in Computational Intelligence" (SCI) publishes new developments and advances in the various areas of computational intelligence—quickly and with a high quality. The intent is to cover the theory, applications, and design methods of computational intelligence, as embedded in the fields of engineering, computer science, physics and life sciences, as well as the methodologies behind them. The series contains monographs, lecture notes and edited volumes in computational intelligence spanning the areas of neural networks, connectionist systems, genetic algorithms, evolutionary computation, artificial intelligence, cellular automata, self-organizing systems, soft computing, fuzzy systems, and hybrid intelligent systems. Of particular value to both the contributors and the readership are the short publication timeframe and the world-wide distribution, which enable both wide and rapid dissemination of research output.

More information about this series at http://www.springer.com/series/10618

Sudipto Chaki · Sujit Ghosal

Modelling and Optimisation of Laser Assisted Oxygen (LASOX) Cutting: A Soft Computing Based Approach

 Springer

Sudipto Chaki
Department of Automobile Engineering
MCKV Institute of Engineering
Howrah, West Bengal, India

Sujit Ghosal
Department of Mechanical Engineering
Netaji Subhas Engineering College
Kolkata, West Bengal, India

ISSN 2191-530X   ISSN 2191-5318   (electronic)
SpringerBriefs in Applied Sciences and Technology
ISSN 2625-3704   ISSN 2625-3712   (electronic)
SpringerBriefs in Computational Intelligence
ISBN 978-3-030-04902-7   ISBN 978-3-030-04903-4   (eBook)
https://doi.org/10.1007/978-3-030-04903-4

Library of Congress Control Number: 2018962383

This Springer imprint is published by the registered company Springer Nature Switzerland AG
The registered company address is: Gewerbestrasse 11, 6330 Cham, Switzerland

Preface

In conventional laser cutting of steel plates, oxygen is used as an assist gas to enhance the process speed and depth of cut through generation of exothermic energy from combustion of iron. But, laser power requirement increases with thickness of steel plates and increases process cost substantially. Laser assisted oxygen cutting (LASOX) has been emerged as a successful method for thick section cutting of mild steel with low power laser. The present book initially detailed the basics of LASOX cutting process, its development, advantages, disadvantages and the details of the research work carried out so far for modelling and optimisation of the process. LASOX cutting process involves complex mathematical relationship among the process variables and soft computing techniques can be suitably employed for modelling and optimisation of such process. Soft computing mimics the remarkable ability of the human mind to reason and learn in an environment of uncertainty and imprecision. It consists of artificial neural networks (ANN), fuzzy set theory, and derivative-free optimisation methods such as genetic algorithm (GA), simulated annealing (SA) etc. However, very little work has been observed on application of soft computing in LASOX cutting process. Present book introduced two integrated soft computing based models (ANN-GA and ANN SA) for modelling and optimisation of the LASOX cutting along with detailed discussion on basic working algorithms of soft computing tools such as ANN, GA, SA etc. Finally, a case study on CO_2 LASOX cutting problem has been carried out where experimental dataset has been employed for ANN modelling and optimisation. Cutting speed, gas pressure, laser power and stand off distance of the process have been considered as controllable input variables and HAZ width, kerf width and surface roughness are considered as output parameters. Three different ANN algorithms such as back propagation neural networks (BPNN) with Levenberg Marquardt, BPNN with Bayesian Regularisation and Radial Basis Function Networks are used for ANN modelling. Best ANN architecture is selected based on prediction/testing performances and integrated with GA and SA to develop integrated ANN-GA and ANN-SA models. Such integration of ANN eliminates the need of closed form objective function during optimisation and a single model completes modelling and optimisation of LASOX cutting process subsequently in a single run. Finally, best optimisation model has been selected on comparing integrated models based on optimisation capability. Significance of the optimised parameter setting has been discussed in detail. The book is not only providing a notion to model and optimize LASOX parameters through soft computing based integrated models but also providing a practical implementation of the proposed model. The authors hope that, the book will be useful to researchers as well as to the practicing engineers working in this field.

Howrah, India
Kolkata, India

Sudipto Chaki
Sujit Ghosal

Contents

About the Authors

Sudipto Chaki received his B.Tech. degree in Mechanical Engineering from North Eastern Regional Institute of Science and Technology, Itanagar, India in 2003, M.Tech. Degree in Manufacturing Technology from National Institute of Technical Teachers' Training and Research, Kolkata, India in 2007 and Ph.D. in engineering from Jadavpur University in 2013. He is presently working as Associate professor and Head in Automobile Engineering Department of MCKV Institute of Engineering, West Bengal, India. He has published 28 research papers so far, in national and international journals and conferences. His research interest includes applications of Soft Computing techniques for modeling and optimisation of manufacturing processes.

Sujit Ghosal received the B.E. degree in Mechanical Engineering from Jadavpur University, India in 1976, M.Tech. Degree in Mechanical Engineering from the same institute in 1988 and Ph.D. in Engineering from Indian Institute of Technology, Kharagpur, India in 1993. He retired as a Professor at the Mechanical Engineering Department of Jadavpur University, India in 2018 and presently working as Professor, Mechanical Engineering Department, Netaji Subhas Engineering College, Garia, Kolkata, India. He has more than 25 years of teaching and research experience and has published large number of research papers in national and international journals and conference proceedings. He is also working as technical advisor of different Government and multi-national organizations. His research interest includes CFD, Droplet Combustion, optimization and Hydraulic Systems.

Chapter 1
LASOX Cutting: Principles and Evolution

Abstract Nowadays, gas-assisted laser cutting of mild and carbon steels has been widely used in manufacturing industries for its accuracy and efficiency. But laser power requirement increases with thickness of the plate and increases process cost substantially. Laser assisted oxygen cutting (LASOX) is an effective method for cutting thick section cutting of mild steel with low power laser. The present chapter has explained the development of the LASOX process with its basic working principle. Researches carried out so far for modelling and optimisations of the LASOX process parameters using conventional statistical methods also have been discussed. But it has been observed that, limitations of statistical methods for modelling of complex non linear relationship between variables of laser material processing can be overcome by incorporating soft computing techniques. Those techniques are widely used in different laser material Processing. Some significant recent works have been discussed in the present chapter. But it has been yet to be incorporated in LASOX processes. Chapter ends with a notion to develop constructive integrated soft computing models for modelling and optimisation of LASOX processes.

Keywords Laser assisted oxygen cutting · Modelling · Optimisation
Soft computing

1.1 Introduction

Emergence of difficult-to-machine advanced engineering materials, stringent design requirements, complex shape and unusual size of work piece has restricted the use of conventional machining methods, in recent years. As a consequence, several non conventional machining techniques known as advanced machining processes (AMPs) have been developed. Laser beam machining (LBM) is one of the most versatile AMPs which can be used for shaping almost the whole range of engineering materials such as, metals and non-metals, soft and difficult-to-machine (DTM) materials. Main source of energy for LBM is Laser, an acronym for light

© The Author(s), under exclusive license to Springer Nature Switzerland AG 2019
S. Chaki and S. Ghosal, *Modelling and Optimisation of Laser Assisted Oxygen (LASOX) Cutting: A Soft Computing Based Approach*, SpringerBriefs in Computational Intelligence, https://doi.org/10.1007/978-3-030-04903-4_1

amplification by stimulated emission of radiation, which is essentially a coherent, convergent and monochromatic beam of electromagnetic radiation with wavelength ranging from ultra-violet to infrared (Steen 2005). The increasing demand of laser in material processing can be attributed to several unique advantages of laser namely, high productivity, automation worthiness, non-contact processing, elimination of finishing operation, reduced processing cost, improved product quality, greater material utilisation and minimum heat affected zone. LBM being a thermal process mainly depends on thermal properties and partially on the optical properties rather than the mechanical properties of the material to be machined. Therefore, both conductive and non-conductive materials can be machined with LBM. Since energy transfer between the laser and the material occurs through irradiation, no cutting forces are generated by the laser and it leads to the absence of mechanically induced material damage, tool wear and machine vibration. Moreover, the material removal rate (MRR) of laser machining is not limited by constraints such as maximum tool force, built up edge formation or tool chatter. In LBM a single machine in combination of a laser beam generating system and multi-axis workpiece positioning system or robot, can be used for drilling, cutting, grooving, welding and heat treating processes. From the application point of view, LBM can be broadly divided into four major categories, namely, forming (manufacturing of near net-shape or finished products), joining (welding, brazing, etc.), machining (cutting, drilling, etc.) and surface engineering (processing confined only to the near-surface region). Laser is also used to perform turning as well as milling operations.

Laser cutting, the most established laser materials processing technology, is a method for shaping and separating a work piece into segments of desired geometry. Laser beam cutting process is a thermal energy based advanced machining process in which the material is removed by (i) melting, (ii) vaporisation, and (iii) chemical degradation (chemical bonds are broken which causes the materials to degrade). When a high energy density laser bream is focused on work surface the thermal energy is absorbed which heats and transforms the work volume into a molten, vaporised or chemically changed state that can easily be removed by flow of high pressure assist gas jet (which accelerates the transformed material and ejects it from machining zone) (Steen 2005). Laser cutting process is dependent on the controllable process variables like laser power, cutting speed, pulse frequency, pulse duration/width, gas pressure, focal length, thickness and composition of work material, type and pressure of assist gas etc. Corresponding relevant output parameters would include material removal rate (MRR), surface roughness, kerf quality, heat affected zone (HAZ) and metallographic properties.

In recent years, gas-assisted laser cutting of mild and carbon steels has been widely used in manufacturing industries for its accuracy and efficiency. It has been observed that, use of oxygen as assist gas enhances the process speed and increases the depth of cut through the release of exothermic energy from the combustion of iron. The process is found to behave well for thin sheets. But laser power requirement increases with the thickness of the section. Several research investigations have been carried out to develop methods for cutting thick sections with low power laser in recent past.

1.1.1   Investigations on Cutting of Thick Sections

In oxygen assisted laser cutting, it had been observed that, the process behaved well during cutting of steel sections up to 15 mm thick with sub-2 kW lasers. But cutting of steel sections having thickness above 15 mm were required laser power substantially above 2 kW with reduction in process stability. Above 25 mm thickness, laser cutting were severely restricted. Several research works had been therefore carried out to develop methods that can cut a section up to 50 mm thickness with a sub-2 kW laser. Fukaya et al. (1990) had demonstrated that a 2 kW CO_2 laser beam can cut up to 20 mm thick mild steel with good surface quality. Molian(1993) developed a dual beam technique involving two CO_2 gas lasers with a power capacity of 1.5 kW each for cutting of steel plates and it has been observed that cutting thickness and speed are enhanced without deteriorating the quality of cut. Kar et al. (1996) had developed a mathematical model for thick section stainless steel cutting with a high power (10 kW) chemical oxygen iodine laser (COIL) and successfully compared with experimental results. Alfille et al. (1996) studied capacity of pulsed Nd:YAG and CO_2 laser for cutting of thick stainless steel plate. Adaptive optics was used to cut sheets of thickness up to 16 mm in mild steel without decrease in cut surface with increase in thickness. Arai et al. (1997) had developed a spinning laser beam method for cutting of thick stainless steel plate with 2 KW laser. Finally, O'Neill and Gabzdyl (2000) had developed Laser assisted oxygen cutting (LASOX), a successful method for thick section cutting of mild steel with low power laser. However, in spite of it being a successful method very little work has been done till date on LASOX cutting method.

1.1.2   Principle of Laser Assisted Oxygen (LASOX) Cutting

In conventional laser cutting process with oxygen as assist gas, the main product of oxy-iron combustion is FeO (97.6%) with Fe_2O_3 and Fe_3O_4. The reaction of $Fe + \frac{1}{2}O_2 \rightarrow FeO$ liberates heat energy of $\Delta H_c = -257.58$ kJ/mol. If it is assumed that, all of the metal removed from a thick section is ignited to form FeO, the total heat of complete combustion is given by $E_r = -4600$ kJ/kg (Chen 1999). During cutting operation, a major part of heat energy of combustion is conducted away from the interaction points to elevate local temperature of the metal far away from laser beam. If the local temperature of the mild steel in this region reaches the ignition point (~ 1000 °C) side burning takes place resulting in wider kerf width and high surface roughness for the cut. The amount of side burning can be reduced with high cutting speed and low gas pressure. However, in conventional laser cutting, a major part of oxidation energy is used for conduction heating and contributes only 25% of the total energy required for cutting. To cut thick metal section with lower laser power, contribution of oxidation energy in cutting is to be increased.

In the process of LASOX cutting, laser beam heat the surface of the steel to the ignition point across the whole area of gas jet impingement. Therefore, the laser power requirements are much reduced and the dynamics of the gas jet control the process output. For complete use of the potential iron-oxygen reaction energy it should be ensured that all of the oxygen striking the surface could engage in the reaction process. This can be achieved by making the laser beam heavily divergent on leaving the nozzle. Figure 1.1 is a schematic of a typical LASOX cutting arrangement. The nozzle assembly has a short focal length lens so that the focal point of the laser beam is situated inside the lens housing and the diverging beam is used to heat metal surface. The co-axial oxygen stream comes out of the nozzle with gas and directed onto the surface of the steel. As the exothermic reaction has a very high yield above 1000 °C, the surface area on which the oxygen jet interacts must be above this temperature for complete combustion within the gas footprint. A heavily divergent beam of moderate power industrial laser is capable of raising the temperature of the steel surface to well above 1000 °C with an area greater than that of the gas jet. Therefore in LASOX cutting, to maximize the exothermic reaction yield, two LASOX conditions are to be achieved, such as,

(i) Divergent Laser beam interaction area should be greater than the gas jet interaction area
(ii) Temperature rise of the steel plate by diverging gas jet should be more than 1000 °C.

If the LASOX condition is achieved, 80% of the exothermic energy will be used for cutting. The reaction process will then propagate in waves down through the depth of the metal and a good cut can be achieved. If the above condition is not met, then intermittent and uncontrolled reaction will occur, with very poor edge quality.

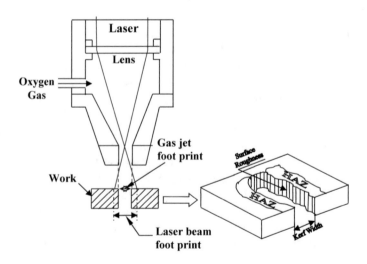

Fig. 1.1 Schematic diagram of LASOX cutting process with measurable output parameters

But on achieving LASOX condition clean dross-free cuts can be attained at speeds comparable to flame cutting using simple nozzles. The ability of the LASOX process combined with the availability of cheaper laser systems may provide a low-cost solution to the thick section cutting problem without having to revert to high cost and high beam quality lasers.

1.1.3   Comparison of LASOX Cutting Process with Conventional Flame Cutting

Form the basic principle it seems that, LASOX cutting is more similar to conventional flame cutting compared to conventional laser cutting. But LASOX cutting have certain advantages compared to flame cutting (O'Neill and Gabzdyl 2000), such as:

1. LASOX cutting requires no pre-heat time, while flame cutting requires several seconds to generate the ignition conditions.
2. When optimised LASOX cutting produces a minimum HAZ, while flame cutting produces significant HAZ on the top of the plate due to the non-localised heat input from the crown of flames.
3. LASOX cutting does not require multiple gases unlike flame cutting.
4. LASOX cutting can pierce profile and trepan as can flame cutting.
5. LASOX cutting does not suffer from top edge melt unlike flame cutting.
6. LASOX cutting can be retrofitted to laser-cutting systems given that sufficient work handling ability exists (thick plates are very heavy and the excessive heat transfer to the cutting bed through high dross volumes may damage work-handling systems).
7. LASOX cutting requires a simple nozzle to operate unlike flame cutting which requires complex nozzle constructions.
8. LASOX cutting requires no extra control functions from the cutting system, as these are supplied by existing systems.
9. LASOX cutting can produce very thin webs without burnout. This provides for a higher density of nested parts. Webs as thin as 3 mm can be produced on 30 mm thick steel.

1.2   Recent Research Trends in Laser Assisted Oxygen (LASOX) Cutting

Initial work of O'Neill and Gabzdyl (2000) on LASOX cutting had used 2.0 kW CO_2 industrial laser for cutting up to 50 mm thick steel plates. In order to ensure the LASOX condition, the laser beam diameter was set to 4 mm at the surface of the steel which was larger than gas jet diameter of 3 mm. It has been shown that, on achieving LASOX condition, clean dross-free cuts can be attained at speeds comparable with flame cutting using simple nozzles.

After establishment of the process it was very much needed a detailed analysis on effect of the LASOX process parameters on cutting quality characteristics. Zaitsev et al. (2007) has performed High-quality LASOX cutting of steel sheets of thickness up to 50 mm through simultaneous action of laser radiation and a supersonic oxygen jet on the material. Parameters of the nozzle used for jet formation are determined by simulating numerically three-dimensional flows of a viscous and heat-conducting gas in a plane channel that is geometrically similar to the laser cut.

Sundar et al. (2009) had carried out an experimental study on LASOX cutting of 40 mm thick mild steel plate with 1 kW laser to investigate influence of the combined effect of the distance between the nozzle and the sheet, cutting velocity, laser power and pressure of oxygen in the settling chamber on the surface roughness, cut width and heat affected zone. Statistical tool like response surface methodology (RSM) has been used for the purpose.

Ermolaev et al. (2013) have studied specific features of hybrid laser-assisted oxygen cutting of mild steel sheets theoretically and experimentally. The shape and geometrical size of a supersonic confuser-diffuser nozzle are demonstrated to play an important role in the formation of oxygen cutting jet. Through numerical solution of three-dimensional Navier–Stokes equations It has been revealed that, a pseudoshock phenomenon in the gas flow in a narrow channel is the reason for elevated roughness in the lower part of the cut under certain conditions. A method of the analytical calculation of effective cutting parameters, which depend on the material thickness, nozzle geometry, and range of oxygen pressure, is proposed in the work. Finally, experimental validation of the proposed formulation has been successfully carried out for cutting steel sheets up to 30–50 mm thick.

But as LASOX cutting parameters bear complex nonlinear relationships among multiple inputs and the output characteristics, it was not easy to construct a mathematical model for employing online process monitoring concept for them. Generally, statistical techniques such as Taguchi Method, Factorial Design and Response Surface Methodology are used to analyse and optimise the process parameters. But they require closed form regression equation as objective function and inherent limitation in accuracy of the formulation of regression equations often leads to inaccurate prediction and optimisation results. Moreover, those processes are not suitable for implementation in online process control. In such condition, soft computing, an innovative approach to construct computationally intelligent systems has emerged as an essential tool for process modelling and optimisation.

1.3   Soft Computing Techniques

Soft computing is an emerging approach to computing which parallels the remarkable ability of the human mind to reason and learn in an environment of uncertainty and imprecision (Zadeh 1992). Soft computing consists of several computing paradigms, including artificial neural networks (ANN), fuzzy set theory, approximate reasoning and derivative-free optimization methods such as genetic algorithm (GA) and simulated annealing (SA). The seamless integration of these methodologies forms the core of soft computing; the synergism allows soft computing to incorporate human knowledge effectively, deal with imprecision and uncertainty, and learns to adapt to unknown or changing environment for better performance.

Soft computing techniques can easily model input output relationship for any complex physical process and thereby can act as an efficient supervisory control for online process monitoring to maintain the product quality in optimised region.

1.3.1   Artificial Neural Networks (ANN)

Artificial neural networks (ANN) are inspired from the structure of the human brain. They are massively parallel adaptive networks consisting of simple nonlinear computing elements called neurons. Each neuron receives signals as input which passes through a weighted pathway in order to generate a linear weighted aggregation of the impinging signals (Haykin 2006). It can be then certainly transformed through an activation function to generate the output signal of the neuron. The activation functions may be binary threshold, linear threshold, Sigmoidal, Gaussian etc. During training phase ANN approximate the underlying functional relationship between input-output variables of a given process to any arbitrary degree of accuracy. During testing phase prediction capability of the trained network is assessed. ANN is generally classified into two categories such as feed forward and recurrent. In Feed Forward Neural Networks (FFNN), information is passed into one direction that is starting from input layer towards the output layer through hidden layer(s). So, it does not form any cycle or loop. Among different variants of Feed Forward Neural Networks, Back Propagation Neural Networks (BPNN) is popular in the field of process modelling of LASOX cutting.

1.3.2   Genetic Algorithm (GA)

Genetic Algorithms (GA) are derivative free stochastic optimisation methods based loosely upon the concept of natural selection and natural genetics (Deb 2005).GA has become popular due to its lesser tendency to get trapped into local minima and is generally expected to find global solution because it works with population instead of a single point as in traditional optimisation method. During GA optimisation, a random population is generated initially, which is further modified over iterations by subsequent operations like Objective Function Evaluation, Fitness Scaling to obtain fitness function value, Selection of above average population members and generation of new population members using Crossover and Mutation. The computation stops if there is no further improvement in the best fitness value for a certain number of consecutive iterations(or reproduction of new generations) cycles. Moreover, GA is applicable to both continuous and discrete optimisation problems. Therefore, it can be suitably employed for optimisation of LASOX processes.

1.3.3 Simulated Annealing (SA)

Simulated Annealing (SA) is a random search technique generally used for function minimisation problem that resembles annealing in metallurgical practice, where the molten metal attains the crystalline state with minimum possible free energy through slow cooling. The cooling phenomenon is simulated by Boltzmann probability distribution. According to Boltzmann probability distribution, a system in thermal equilibrium at a temperature T has its energy distributed probabilistically as $P(E_n) = \exp(-E_n/kT)$, where k is the Boltzmann constant or cooling rate. It indicates that a system at high temperature has almost uniform probability of being at any energy state but at low temperature it has small probability of being at high energy state. Simulated Annealing algorithm uses this concept of minimum energy by representing energy E_n with the objective function value with a control parameter likened to the system temperature T (Deb 2005).

1.3.4 Some Significant Contributions on Application of Soft Computing Techniques in Laser Material Processing

Soft computing techniques have been widely used for modelling and optimisation of different laser material processing processes in recent years. Some of the significant works are as follows:

Luo et al. (2005) had used standard BPNN with adaptive learning rate and momentum to diagnose welding faults during CO_2 laser welding using both keyhole and conduction modes. Features extracted from the acoustic signals were input into the ANN so that after training, the ANN could be used to identify between normal and abnormal welds.

Olabi et al. (2006) had employed an integrated approach of BPNN and TM for optimisation of penetration-to-fuse-zone-width and the penetration-to-HAZ-width ratios, during CO_2 keyhole laser welding of medium carbon steel butt weld where welding speed, laser power and focal position are considered as input parameters.

Yilbas et al. (2008) had employed BPNN to classify the dross height and out of flatness of the cut surfaces for CO_2 laser cutting of stainless steel for the inclined surfaces and normal surfaces. It was found that the dross height and out of flatness were influenced significantly by the laser output power, particularly for wedge-cutting situation.

Tsai et al. (2008) had employed a BPNN model with Levenberg–Marquardt (LM) algorithm for prediction of cutting qualities such as depths of the cutting line, widths of HAZ and cutting line during cutting of Quad Flat Non-lead (QFN) packages by using a Diode Pumped Solid State Laser (DPSSL) System. Finally, a genetic algorithm (GA) was applied with user defined objective function to find the optimal cutting parameters that lead to least HAZ width and fast cutting speed with complete cutting.

Park and Rhee (2008) had studied tensile strength of AA5182 aluminum alloy during Nd:YAG laser butt welding with AA5356 filler wire where laser power, welding speed and wire feed rate are considered as input parameters. Using the experimental results, a BPNN model with LM algorithm had been trained to predict the tensile strength. Further, a fitness function for GA had been formulated, taking into account weldability and productivity while the laser power, welding speed and wire feed rate were optimised. The optimal welding conditions was achieved with high tensile strength, low wire feed rate and fast welding speed.

Canyurt et al. (2008) had developed a genetic algorithm laser welding strength estimation model (GALWSEM) for estimation of welding strength during Nd:YAG laser-MIG hybrid welding of 6K21-T4 aluminium alloy. Input process parameters are wire type, shielding gas, laser energy, laser focus, travelling speed and wire feed rate. A regression model developed for connecting input parameters with welding strength has been employed as objective function for GALWSEM. The results indicate that, tensile strength initially decreases and then increases with the increment of laser power.

Kuo et al. (2011) had employed BPNN and the Levenberg–Marquardt (LM) algorithm were integrated to establish a prediction system to monitor laser material processing of a polymethyl methacrylate (PMMA) light guide plate, a part of a back light module, using a CO_2 laser.

Kadri et al. (2015) have compared the performance of an artificial neural network (ANN) and Finite Element (FE) model to predict thermal stresses at the leading and trailing edges of the glass sheet during its cutting using diode laser with varying thickness and laser cutting speed. Results indicate that, predicted results from ANN are better than the FE model.

1.3.4.1   Trends of Application of Soft Computing Techniques in Laser Material Processing

The literature indicates a general trend to employ ANN for prediction of output quality parameters. In all ANN applications BPNN with either gradient descent momentum or with LM is used as training algorithm. ANN has been shown better prediction capability compared to regression models and FE models. GA is used for optimisation of different LBM processes including laser cutting, welding etc. where a separate regression equation is developed and used as fitness function. Though some works have included ANN and GA, they have treated prediction and optimisation separately. GA optimisation is completed with separate regression equation with no influence of ANN.

1.3.4.2   Application of Soft Computing Techniques in LASOX Cutting

So far scanty literature is available regarding soft computing modeling and optimization of LASOX cutting process parameters. Chaki and Ghosal (2011) have developed an optimized SA-ANN model of artificial neural network (ANN) and simulated annealing (SA) to predict and optimize cutting quality of LASOX cutting process of mild steel plates. Optimization of SA-ANN parameters is carried out first where the ANN architecture and initial temperature for SA are optimized. The optimized ANN architecture is further trained using single hidden layer back propagation neural network (BPNN) with Bayesian regularization (BR). The trained ANN is then used to evaluate the objective function during optimization with SA. Experimental dataset employed for the purpose consists of input cutting parameters comprising laser power, cutting speed, gas pressure and stand-off distance while the resulting cutting quality is represented by heat affected zone (HAZ) width, kerf width and surface roughness. Experimental validation of the proposed SA-ANN model indicated reasonably good accuracy (around 3%). No other significant work has been observed in the literature regarding soft computing modeling of LASOX cutting processes.

1.4   Soft Computing Modeling and Optimization of LASOX Cutting—An Overview

The present book addresses application of soft computing techniques for LASOX cutting in a synchronized manner. An integrated model or algorithm combining ANN and evolutionary optimisation techniques like GA and SA for modelling and optimisation of LASOX cutting process has been proposed in the following Chapters. In that model initially different ANN algorithms have been employed for approximating the underlying functional relationship between input-output variables of LASOX process. The ANN producing best prediction accuracy is integrated with GA and SA for optimisation of LASOX parameters. That integration of ANN with GA and SA eliminates the need of any closed form objective function. That integrated model can be implemented for online process monitoring.

Chapter 2 of the book will discuss on basic working algorithms of soft computing tools such as ANN, GA, SA etc. It will introduce integrated ANN and evolutionary optimisation algorithm based models for modelling and optimisation of the LASOX cutting. Working of such models will be explained in detail. Chapter 3 will carry out a case study on a LASOX cutting problem. Here, process parameters of an LASOX cutting experimental dataset will be modelled and optimised through the proposed model. Finally, the major conclusions will be discussed.

References

Alfille JP, Pilot G, De Prunele D (1996) New pulsed YAG laser performances in cutting thick metallic materials for nuclear applications. In: Proceedings of SPIE, pp 134–144

Arai T, Riches S (1997) Thick plate cutting with spinning laser beam. Laser Inst Am 83(1):19–26

Canyurt OE, Kim HR, Lee KY (2008) Estimation of laser hybrid welded joint strength by using genetic algorithm approach. Mech Mater 40(10):825–831

Chaki S, Ghosal S (2011) Application of an optimised SA-ANN hybrid model for parametric modelling and optimisation of LASOX cutting of mild steel. Prod Eng Res Dev 5(3):251–262 (Springer, Heidelberg)

Chen SL (1999) The effects of high pressure assistant gas flow on high power CO_2 laser cutting. J Mater Process Technol 88(1–3):57–66

Deb K (2005) Optimisation for engineering design: algorithms and examples, 8th edn. Prentice-Hall of India Private Limited, India

Ermolaev GV, Kovalev OB, Zaitsev AV (2013) Parameterization of hybrid laser-assisted oxygen cutting of thick steel plate. Opt Laser Technol 47:95–101

Fukaya K, Karube N (1990) Analysis of CO_2 laser beam suitable for thick metal cutting. Laser Inst Am 71:61–70

Haykin S (2006) Neural networks: a comprehensive foundation, 2nd edn. Pearson Education Inc, India

Kadri MB, Nisar S, Khan SZ, Khan WA (2015) Comparison of ANN and Finite Element Model for the prediction of thermal stresses in diode laser cutting of float glass. Optik—Int J Light Electron Opt 126(19):1959–1964

Kar A, Scott JE, Latham WP (1996) Theoretical and experimental studies of thick-section cutting with a chemical oxygen-iodine laser (COIL). J Laser Appl 8:125–133

Kuo C-FJ, Tsai W-L, Su T-L, Chen J-L (2011) Application of an LM-neural network for establishing a prediction system of quality characteristics for the LGP manufactured by CO_2 laser. Opt Laser Technol 43(3):529–536

Luo H, Zeng H, Hu L, Hu X, Zhou Z (2005) Application of artificial neural network in laser welding defect diagnosis. J Mater Process Technol 170(1–2):403–411

Molian PA (1993) Dual-beam CO_2 laser cutting of thick metallic materials. J Mater Sci 28:1738–1748

Neill WO, Gabzdyl JT (2000) New developments in oxygen-assisted laser cutting. J Opt Lasers Eng 34(4–6):355–367

Olabi AG, Casalino G, Benyounis KY, Hashmi MSJ (2006) An ANN and Taguchi algorithms integrated approach to the optimization of CO_2 laser welding. Adv Eng Softw 37(10):643–648

Park YW, Rhee S (2008) Process modeling and parameter optimization using neural network and genetic algorithms for aluminum laser welding automation. Int J Adv Manuf Technol 37:1014–1021

Steen WM (2005) Laser material processing, 3rd edn. Springer, London

Sundar M, Nath AK, Bandyopadhyay DK, Chaudhuri SP, Dey PK, Misra D (2009) Effect of process parameters on the cutting quality in Lasox cutting of mild steel. Int J Adv Manuf Technol 40(9–10):865–874

Tsai M-J, Li C-H, Chen C-C (2008) Optimal laser-cutting parameters for QFN packages by utilizing artificial neural networks and genetic algorithm. J Mater Process Technol 208(1–3):270–283

Yilbas BS, Karatas C, Uslan I, Keles O, Usta Y, Yilbas Z, Ahsan M (2008) Wedge cutting of mild steel by CO_2 laser and cut-quality assessment in relation to normal cutting. Opt Lasers Eng 46 (10):777–784

Zadeh LA (1992) Fuzzy logic, neural networks and soft computing, one-page course announcement of CS 294-4, spring 1993. University of California, Berkley

Zaitsev AV, Kovalev OB, Malikov AG, Orishich AM, Shulyat'ev VB (2007) Laser cutting of thick steel sheets using supersonic oxygen jets. Quantum Electron 37:891–892

Chapter 2
Integrated Soft Computing Based Methodologies for Modelling and Optimisation

Abstract The chapter has explained working of two integrated soft computing based models comprising of Artificial Neural Networks (ANN), Genetic Algorithms (GA) and Simulated Annealing (SA) namely, ANN-GA and ANN-SA for modelling and optimisation of Laser Assisted Oxygen Cutting (LASOX) process. Three different ANN training algorithms such as back propagation neural networks (BPNN) with Levenberg Marquardt (LM) algorithm, BPNN with Bayesian Regularisation (BR) and Radial Basis Function Networks (RBFN) have been employed for training and subsequent prediction. The ANN producing best prediction performance is used for prediction of objective function value. That process is thereby eliminating the need for closed form objective functions to produce required optimisation accuracy. The Chapter detailed the working of ANN, GA, SA and integrated models in step by step for easy understanding of the readers. The process has been explained in a generalised manner so that it can be applied for modelling and optimisation of any manufacturing process.

Keywords Artificial neural networks(ANN) · Back propagation neural networks Levenberg marquardt algorithm · Bayesian regularisation · Radial basis function networks · Genetic algorithms (GA) · Simulated annealing (SA) Integrated ANN-GA · Integrated ANN-SA

2.1 Introduction

Integrated soft computing based approaches for process modelling and optimisation of LASOX cutting have been proposed in the present chapter. Program codes written in MATLAB 7.0 environment are used for two-way integration of prediction and optimisation modules. Experimental dataset employed for formulation and validation of the models is divided into training and testing dataset for the purpose of ANN training and testing for prediction capability. Different architectures for each of three different ANN training algorithms such as BPNN with LM, BPNN with BR and RBFN have been employed for training and subsequent prediction.

Then best ANN is selected on comparing prediction performance of all tested architectures. Best ANN architecture is further compared with conventional regression model to check possible superiority of ANN over conventional regression models. Finally, best ANN architecture is integrated with GA and SA to develop integrated ANN-GA and ANN-SA models where the trained ANN is employed for calculation of objective function value. Integration of ANN eliminates the need of closed form objective function. Further, optimisation performance of ANN-GA and ANN-SA is compared to determine the most effective model for LASOX processes. The steps of the methodology adopted are provided in Fig. 2.1. Working of ANN, regression model, GA and SA optimisation technique with their combination have been explained in the subsequent sections. The same models can be successfully used for modelling of any other manufacturing processes.

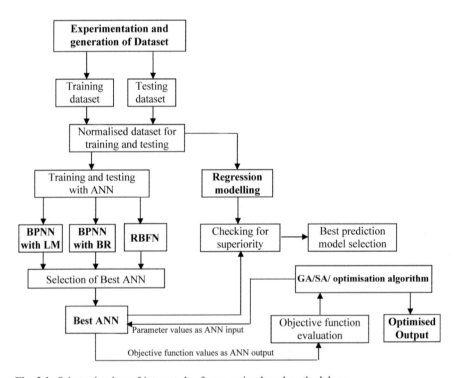

Fig. 2.1 Schematic view of integrated soft computing based methodology

2.2   Prediction Modelling with Artificial Neural Networks

Artificial Neural Networks (ANN) are massively parallel adaptive networks of simple nonlinear computing elements called neurons which are intended to abstract and model some of the functionality of the human nervous system in an attempt to partially capture some of its computational strengths. This concept of ANN is an outcome of the working of biological neurons. Some properties that make artificial neural network popular for real world applications are robustness, capacity for associative recall and capability for function approximation and generalisation. Prediction capability of ANN is dependent on its training performance. Training/ Learning is generally of two types e.g. supervised and unsupervised. In supervised learning, training is performed using a knowledge base i.e. a given input-output dataset to develop underlying functional relationship through minimisation of prediction errors between ANN predicted value and desired output values. In unsupervised learning, as desired output data are not available, prediction error cannot be determined and network passes through a self organising process to classify input data in groups. In the present model, supervised learning of feed forward ANN has been used. Multilayer Feed Forward ANN, also known as 'Universal Function Approximator' (Hornik, Tinchcombe and White 1989) is attractive to the research community for such parametric modelling due to its inherent capability to approximate the underlying function of a given data set to any arbitrary degree of accuracy. On successful completion of training, a testing phase is carried out to assess prediction capability of the trained network using test input dataset that was not used during training.

In the present model, two types of feed forward neural networks i.e. BPNN and RBFN have been used for prediction of quality parameters during LASOX cutting for a set of input process parameters. Two different schemes of BPNN, namely, BPNN with Levenberg-Marquardt (LM) and BPNN with Bayesian Regularisation (BR) have been considered. Performance of two BPNN models and RBFN model are compared for selection of the final algorithm.

A single program is developed for repeated training and subsequent testing of a number of ANN architectures in MATLAB 7.0 environment. In the program for training and testing with different training algorithms, it is only required to call the specific subroutines in the main program. Training and testing data of ANN is stored in a data file. Working of the schemes employed is detailed in the following sections.

2.2.1   Working of Back Propagation Neural Network (BPNN)

Initial steps are similar for both BPNN with LM and BPNN with BR. Variation in two training algorithms are observed only in the methodologies for updating weights. Workings of both the algorithms are given below:

2.2.1.1 Setting the Training Input and Output Data

Present work has employed single hidden layer ANN architecture that typically consists of one input layer, one hidden layer and one output layer as shown in Fig. 2.2. Input and output layers consist of input parameters and output responses for the experiments. Naturally, number of neurons in input and output layer is fixed, each being equal to the no. of input parameters and output responses, respectively. If in an experiment there are m number of experimental input parameters and p number of output responses desired, the input and output vectors are given by,

$$\mathbf{X} = [X_1 \quad X_2 \quad \ldots\ldots\ldots \quad X_m], \quad \mathbf{D} = [D_1 \quad D_2 \quad \ldots\ldots\ldots \quad D_p] \quad (2.1)$$

Thus, the no. of input layer neurons is m and that for the output layer is p. Number of hidden layer neurons is generally considered as variable. Number of neurons in hidden layer is gradually increased to find optimum hidden layer neurons for best training/prediction performance.

Number of hidden layer neurons is generally considered as variable. All the neurons in input, hidden and output layers bear weighted connections. Network formation is initiated with random distribution of weights which are updated over epochs to obtain final distribution of weights. At this step of BPNN with BR training, value of relevant regularisation parameters, α and β, along with other network parameters like error tolerance etc. are also initialised while during training for BPNN with LM, values for scaling coefficient $\mu(\mu \geq 0)$, decay rate δ $(0 < \delta < 1)$, error tolerance etc. are initialised.

2.2.1.2 Normalisation of Data

In order to increase accuracy and speed of the network, input and output data are normalised between 0 and 1 before actual application in the network. Using the

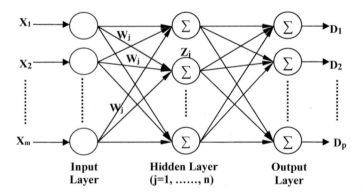

Fig. 2.2 Architecture of the Back propagation Networks

maximum and minimum real values of input $(\mathbf{X}_{max}, \mathbf{X}_{min})$ and maximum real values of output $(\mathbf{D}_{max})$, normalised input $(\mathbf{X}_{nor})$ and output $(\mathbf{D}_{nor})$ are computed as follows:

$$\mathbf{X}_{nor} = \frac{\mathbf{X} - \mathbf{X}_{min}}{\mathbf{X}_{max} - \mathbf{X}_{min}}, \quad \mathbf{D}_{nor} = \frac{\mathbf{D}}{\mathbf{D}_{max}} \tag{2.2}$$

Normalisation of output as in Eq. (2.2) has shown less training error.

80% of the normalised input $(\mathbf{X}_{nor})$ and output $(\mathbf{D}_{nor})$ data are chosen randomly as training input $(\mathbf{X}_{nor}^{Train})$ and output $(\mathbf{D}_{nor}^{Train})$ data while remaining 20% data are reserved as testing input $(\mathbf{X}_{nor}^{Test})$ and output $(\mathbf{D}_{nor}^{Test})$ data.

2.2.1.3   Forward Pass Computation

Normalised input data is multiplied with corresponding initial (and subsequently updated) weights to produce activation values of hidden layer neurons. For the jth neuron of the hidden layer, activation value (M_j) is calculated as,

$$M_j = \mathbf{w}_j^T \mathbf{X}_{nor}^{Train}, \quad \text{for } j = 1, 2, \ldots\ldots\ldots, n, \tag{2.3}$$

where, weight vector $\mathbf{w}_j$ denotes the association between inputs and jth node of the hidden layer and can be defined as, $\mathbf{W}_j \underline{\Delta} [\, W_{j1} \quad W_{j2} \quad \cdots\cdots\cdots \quad W_{jm}\,]$, n is the number of hidden layer neurons. M_j is further converted into output signal Z_j of the jth hidden neuron through sigmoidal activation function and Z_j is given by,

$$Z_j = f(M_j) = \frac{1}{1 + e^{-M_j}} \tag{2.4}$$

Output signals (Z) of all hidden layer neurons are computed in this way and are employed as input to output layers neurons. All Z's are further multiplied with corresponding weights between hidden and output layer and converted into output signal (**O**) of the network through linear activation function.

2.2.1.4   Error Computation, Updating Training Parameters and Convergence

Difference between network output **O** and desired normalised output value $\mathbf{D}_{nor}^{Train}$ is considered as error which is to be minimized. Here, mean squared error (MSE) is considered as performance function for a network and objective function of training process is to minimise the training MSE. It is calculated as,

$$\mathbf{E}_r = \text{Training MSE} = \frac{1}{Q} \sum_{i=1}^{Q} (\mathbf{D}_{\text{nor }\{i\}}^{\text{Train}} - \mathbf{O}_i)^2 \qquad (2.5)$$

where, Q is total number of training data. If computed MSE satisfies the convergence criterion, training stops. Otherwise the calculated error E_r is used to compute the change in network weights such that a global error measure gets reduced over iterations and it is known as back propagation of errors. In conventional gradient descent learning weight changes are made in proportion to the negative of the error gradient $\left(\frac{\partial \mathbf{E}_r}{\partial \mathbf{w}}\right)$. If $\mathbf{w}^k$ is the weight at kth iteration and $\Delta \mathbf{w}^k$ denotes the weight gradient, weight in (k + 1)th iteration can be updated in the general BPNN in following way:

$$\mathbf{w}^{k+1} = \mathbf{w}^k + \Delta \mathbf{w}^k = \mathbf{w}^k + \eta \times \left(-\frac{\partial \mathbf{E}_r}{\partial \mathbf{w}^k}\right) = \mathbf{w}^k + \eta \times (-\nabla \mathbf{E}_r) \qquad (2.6)$$

where, η is considered as learning rate parameter which can be varied according to requirement. A minimum value of η will maintain a smooth trajectory in weight space while a larger η can lead to oscillations during learning. Further, in order to increase rate of learning a momentum term (α) is introduced into the weight update procedure and the weight is updated using the following equation:

$$\mathbf{w}^{k+1} = \mathbf{w}^k + \Delta \mathbf{w}^k = \mathbf{w}^k + \eta \times (-\nabla \mathbf{E}_r) + \alpha \times \Delta \mathbf{w}^{k-1} \qquad (2.7)$$

However, momentum term does not always seem to speed up training, as it is more or less application dependent and is found too slow for many practical problems. Moreover, in traditional gradient descent BPNN weight gradient is computed by first order error gradient which shows inherent limitation in modelling dataset having high nonlinearity.

CASE I: BPNN with LM:
BPNN with Levenberg-Marquardt (LM) algorithm is much faster and can model high nonlinearity in dataset compared to traditional gradient descent algorithm as it can compute up to second order error gradient but eliminating the need to compute the Hessian matrix H (Hagan and Menhaj 1994) during back propagation of errors.

If $\mathbf{w}^t$ be the weight at tth iteration and $\Delta \mathbf{w}^t$, the weight gradient, weight in (t + 1)th iteration can be updated in the following way:

$$\mathbf{w}^{t+1} = \mathbf{w}^t + \Delta \mathbf{w}^t = \mathbf{w}^t - (\mathbf{H} + \mu \mathbf{I})^{-1} \mathbf{J}^T \mathbf{E}_r = \mathbf{w}^t - (\mathbf{J}^T \mathbf{J} + \mu \mathbf{I})^{-1} \mathbf{J}^T \mathbf{E}_r \qquad (2.8)$$

where,

I   is the unit matrix, μ is the scaling coefficient,
$\mathbf{E}_r^t$ is the mean squared error and
J   is the Jacobian of error function

MSE is recomputed in $(t + 1)$th iteration and denoted by $\mathbf{E}_r^{t+1}$. MSE in tth iteration is denoted by $\mathbf{E}_r^t$. For $\mathbf{E}_r^{t+1} < \mathbf{E}_r^t$ μ is updated as $\mu = \mu \cdot \delta$ and control moves to forward pass computation step for next iteration. Otherwise, μ is updated as $\mu = \mu/\delta$ and control returns back to weight computation for further improvement.

CASE II: BPNN with BR:
In order to improve generalisation (or prediction) capability, BR minimizes Φ, which is a linear combination of sum of the squared errors (SSE) and sum of the squared weights (SSW) in the following way:

$$\Phi = \beta \times SSE + \alpha \times SSW \tag{2.9}$$

where SSE is given by, $SSE = \sum_{i=1}^{Q} (\mathbf{D}_{nor\,\{i\}}^{Train} - \mathbf{O}_i)^2$ and regularisation parameters α and β are represented as, $\alpha = \frac{\gamma}{2 \times SSE}$ and $\beta = \frac{q-\gamma}{2 \times SSW}$. Here, q is the number of training datasets and γ denotes effective number of parameters required to reduce error function. In any iteration, Φ is used to compute the Hessian matrix $(\mathbf{H})$ in the following way:

$$\mathbf{H} = \nabla^2 \Phi \approx 2\beta \times \mathbf{J}^T \mathbf{J} + 2\alpha \times \mathbf{I} \tag{2.10}$$

where, $\mathbf{J}$ is the Jacobian of the training set errors and $\mathbf{I}$ is the Identity Matrix. Weights are updated for next iteration using the Hessian $(\mathbf{H})$ in the same way as done by the traditional BPNN with LM algorithm (Eq. 2.8). In the process, regularisation parameters α and β are also updated for next iteration. With updated training parameters, program control returns to *Forward Pass Computation* step for computation of error term, Φ, in the next iteration. The process continues till convergence criterion is met.

As in the present model, performance of different ANN architecture during training with different algorithms have been compared and analysed, performance indices for training algorithms should be similar for ease of comparison. In BPNN with LM training, MSE is the performance index. Therefore, in BPNN with BR training, though convergence is achieved by combined objective function Φ, MSE is computed at the point of convergence and considered as performance index for further analysis in the present study.

2.2.1.5 Testing of Trained ANN

Once the network is trained, in order to test generalisation/prediction capability (Haykin 2006), the trained network is further fed with test input dataset $\left(\mathbf{X}_{nor}^{Test}\right)$ that was not used (i.e. unknown to ANN) during training. The resulting ANN predicted output $(\mathbf{O}^{Test})$ is compared with corresponding known experimental test output

$\left(\mathbf{D}_{\mathrm{nor}}^{\mathrm{Test}}\right)$ to determine the prediction error and is measured by the quantity called testing MSE. It is calculated by,

$$\text{Testing MSE} = \frac{1}{N} \sum_{i=1}^{N} (\mathbf{D}_{\mathrm{nor}\,\{i\}}^{\mathrm{Test}} - \mathbf{O}_{i}^{\mathrm{Test}})^2 \tag{2.11}$$

where, N is the number of testing data. Prediction capability of trained ANN is also computed by mean absolute % error during prediction (or testing) and is given as follows,

$$\text{Mean Absolute \% Error in Prediction} = \frac{1}{N} \sum_{i=1}^{N} \left| \frac{\mathbf{D}_{\mathrm{nor}\,\{i\}}^{\mathrm{Test}} - \mathbf{O}_{i}^{\mathrm{Test}}}{\mathbf{D}_{\mathrm{nor}\,\{i\}}^{\mathrm{Test}}} \times 100 \right| \tag{2.12}$$

During testing with test data, absolute % error in prediction obtained for every test data after corresponding de-normalisation can be assessed by

$$\text{Absolute \% Error in Prediction} = \left| \frac{\mathbf{D}_{i}^{\mathrm{Test}} - \mathbf{O}_{i}^{\mathrm{Test}}}{\mathbf{D}_{i}^{\mathrm{Test}}} \times 100 \right| \tag{2.13}$$

2.2.1.6   Parameter Setting for BPNN Training

The combined process of training and testing is carried out for different architectures of network by varying the number of hidden layer neurons, as a design parameter of the network. All weights are initialised through generation of random numbers before training. For all the cases activation function of the hidden layer and output layer is sigmoidal and linear respectively. Each network is trained over 1000 epochs in this purpose.

During training and testing using BPNN with Bayesian Regularisation (BR), all the above mentioned settings are employed. In addition, initial values of Bayesian regularisation parameters α and β are set at 0 and 1 respectively.

2.2.2   Working of Radial Basis Function Networks (RBFN)

The present model has also employed another class of feed forward neural network known as radial basis function network (RBFN) that computes hidden neuron activations using exponential of a measure of distance (usually Euclidean distance) between the input vector and a prototype vector characterising the signal function at hidden neuron, instead of employing an inner product between the input vector and the weight vector (Eq. 2.3) as in BPNN. There are no weighted connections existing between inputs to hidden layer. However, hidden layer output bears a weighted connection with output. RBFN was originally introduced for the purpose of interpolation of the data points. In the present model, RBFN is used as an exact interpolator where number of basis functions is equal to the input vector dimension. Working of a simple RBFN with a single output is explained below:

Step I: Dataset is normalised using Eq. 2.2. Therefore, the training dataset of input–output pairs for the problem is given by, $\tau = \left\{ \left(\mathbf{X}_{\text{nor } i}^{\text{Train}}, \mathbf{D}_{\text{nor } i}^{\text{Train}} \right) \right\}_{i=1}^{Q}$ where $\mathbf{X}_{\text{nor}}^{\text{Train}}$ and $\mathbf{D}_{\text{nor}}^{\text{Train}}$ indicate training input and output data while Q is the total number of training data. RBFN as an exact interpolator (Kumar 2004) involves search of a map f that takes each input $\mathbf{X}_{\text{nor } i}^{\text{Train}}$ and maps it exactly onto the desired output $\mathbf{D}_{\text{nor } i}^{\text{Train}}$ and is given by

$$f(\mathbf{X}_{\text{nor } \{i\}}^{\text{Train}}) = \mathbf{D}_{\text{nor } \{i\}}^{\text{Train}}, i = 1, 2, \ldots\ldots, Q \tag{2.14}$$

Step II: In RBFN, number of neurons in hidden layer is equal to the number of training data. Therefore, there are Q numbers of hidden layer neurons, where every neuron assumes an activation function or basis function $\varphi\left(\left\| \mathbf{X}_{\text{nor}}^{\text{Train}} - \mathbf{X}_{\text{nor } \{i\}}^{\text{Train}} \right\| \right)$ that computes the Euclidean distance $\left\| \mathbf{X}_{\text{nor}}^{\text{Train}} - \mathbf{X}_{\text{nor } \{i\}}^{\text{Train}} \right\|$ between the applied input $\mathbf{X}_{\text{nor}}^{\text{Train}}$ and a training data point $\mathbf{X}_{\text{nor } \{i\}}^{\text{Train}}$. RBFN generally employs Gaussian basis function as activation function at hidden layer and is computed as

$$\varphi\left(\mathbf{X}_{\text{nor}}^{\text{Train}} \right) = \exp\left(-\frac{\left\| \mathbf{X}_{\text{nor}}^{\text{Train}} - \mu \right\|^{2}}{2\sigma^{2}} \right) \tag{2.15}$$

with a center μ, which may be the training data points in this case and σ is the spread factor which has a direct effect on the smoothness of the interpolating function. Spread factor is used as design parameters and varied to obtain different RBFN architecture.

Step III: The linearly activated output layer bears a weighted connection with hidden layer and produce output as scalar product of hidden layer output and weight vector. Output is computed by a weighted superposition of these basis functions for the output layer and is given by,

$$f\left(\mathbf{X}_{\text{nor}}^{\text{Train}} \right) = \sum_{i=1}^{Q} \mathbf{w}_{i} \varphi\left(\left\| \mathbf{X}_{\text{nor}}^{\text{Train}} - \mathbf{X}_{\text{nor } i}^{\text{Train}} \right\| \right) \tag{2.16}$$

Step IV: Substituting the conditions for exact interpolation from Eq. 2.14 into Eq. 2.16 yields,

$$\sum_{i=1}^{Q} \mathbf{w}_{i} \varphi\left(\left\| \mathbf{X}_{\text{nor } \{k\}}^{\text{Train}} - \mathbf{X}_{\text{nor } \{i\}}^{\text{Train}} \right\| \right) = \mathbf{D}_{\text{nor } \{k\}}^{\text{Train}}, k = 1, \ldots\ldots\ldots, Q \tag{2.17}$$

Upon recasting this system of equations in matrix form, the following definitions are introduced:

$$\mathbf{D}_{\text{nor}}^{\text{Train}} = \left(\mathbf{D}_{\text{nor}\{1\}}^{\text{Train}}, \ldots \ldots \ldots, \mathbf{D}_{\text{nor}\{Q\}}^{\text{Train}} \right)^T$$

$$\mathbf{W} = (w_1, \ldots, w_Q)^T$$

$$\boldsymbol{\Phi} = \begin{pmatrix} \varphi\left(\left\| \mathbf{X}_{\text{nor}\{1\}}^{\text{Train}} - \mathbf{X}_{\text{nor}\{1\}}^{\text{Train}} \right\| \right) & \cdots & \varphi\left(\left\| \mathbf{X}_{\text{nor}\{1\}}^{\text{Train}} - \mathbf{X}_{\text{nor}\{Q\}}^{\text{Train}} \right\| \right) \\ \vdots & \ddots & \vdots \\ \varphi\left(\left\| \mathbf{X}_{\text{nor}\{Q\}}^{\text{Train}} - \mathbf{X}_{\text{nor}\{1\}}^{\text{Train}} \right\| \right) & \cdots & \varphi\left(\left\| \mathbf{X}_{\text{nor}\{Q\}}^{\text{Train}} - \mathbf{X}_{\text{nor}\{Q\}}^{\text{Train}} \right\| \right) \end{pmatrix}$$

The equation for $\mathbf{D}$ can now be written as,

$$\mathbf{D} = \boldsymbol{\Phi}^T \mathbf{W} = \boldsymbol{\Phi} \mathbf{W} \tag{2.18}$$

Since $\boldsymbol{\Phi}$ (Q X Q matrix) is symmetric, weights between hidden and output layer are calculated by inverting the Eq. 2.18, such as

$$\mathbf{W} = \boldsymbol{\Phi}^{-1} \mathbf{D} \tag{2.19}$$

After calculating the weights, the network is in a position to act as interpolator and will predict output (**O**) for any set of input data. Testing of different RBFN architecture have been carried out in the same way as it has been described in Sect. 2.2.1.5.

Prediction capability of different RBFN architecture as obtained from varying spread factors is assessed in the present model through Testing MSE (Eq. 2.11) and Absolute % Error in prediction (Eq. 2.13). Best ANN is obtained on comparing prediction performance of different architectures as well as training algorithms.

2.3 Regression Model

Statistical regression analysis is a potential tool for modelling a process. It can provide a relationship between the input and the output parameters based on experimental results. In this model, a multi-variable regression model with both linear and second order equations has been developed to find out a relationship between the input process parameters and measured outputs for different processes taken up for the present study.

The general linear regression model is given by,

$$Y_u = \sum_{i=1}^{n} b_0 X_{iu} + \sum_{i<j}^{n} b_{ij} X_{iu} X_{ju} + e_u \tag{2.20}$$

The general second order regression model is given by,

$$Y_u = \sum_{i=1}^{n} b_0 X_{iu} + \sum_{i=1}^{n} b_{ii} X_{iu}^2 + \sum_{i<j}^{n} b_{ij} X_{iu} X_{ju} + e_u \qquad (2.21)$$

where Y_u = Responses, X_{iu} = value of ith input process parameter of uth experiment, n = number of input process parameter, b_0, b_{ii}, b_{ij} = second order regression coefficients, e_u = experimental error of uth observation. In the present study, regression model has been developed by using MINITAB 15 Software, commercially available software for statistical analysis. Adequacy of the model has been assessed in the present model through regression coefficient (R-square) value of the regression equations. Near-Unity R-square values indicate good accuracy of fit of the equations of regression model.

Predicted output (**O**) for a set of input is compared with desired experimental output (**D**) and prediction performance of the model developed is measured by Testing MSE (Eq. 2.11) and Absolute % Error in Prediction is calculated as follows,

Absolute % error in Prediction

$$= \left| \frac{\text{Experimental output} - \text{Predicted output of Regression Model}}{\text{Experimental output}} \times 100 \right| \qquad (2.22)$$

Prediction performance of regression model is compared with the best ANN model and the model with best prediction capability is used for further analysis of optimisation.

2.4 Objective Function Formulation

Present model employs optimisation module after completion of ANN training and testing. As experiments generally involve multiple input parameters and output responses, the optimisation is necessarily being multi-objective in nature. The multi-objective optimisation problem can be formulated as:

The objective functions:

Minimise/maximise:

$$D_1(X_1, X_2, .., X_m)$$
$$D_2(X_1, X_2, .., X_m)$$
$$\dots\dots\dots\dots\dots\dots\dots$$
$$D_p(X_1, X_2, .., X_m)$$

Subject to the <u>constraints</u>:

$$X_1^{min} \leq X_1 \leq X_1^{max}, X_2^{min} \leq X_2 \leq X_2^{max}, \ldots \ldots \ldots \ldots \ldots, X_m^{min} \leq X_m \leq X_m^{max} \quad (2.23)$$

In present model, the multi-objective optimisation problem has been converted into a single objective optimization problem using an objective function $\mathbf{J}$ as a weighted sum of the output parameters. Therefore, the optimisation problem is formulated as,

Minimise:

$$\mathbf{J} = \sum_{i=1}^{p} D_i \times W_i \quad (2.24)$$

where,

W_i represent the arbitrarily chosen weight parameters attached to each output

In the present model, during optimisation with integrated ANN-GA and ANN-SA techniques objective function is computed from the outputs, as evaluated by the trained ANN based on the inputs of the optimisation routines.

2.5   Optimisation Using Genetic Algorithm

Genetic Algorithms (GA) are computerized search and optimisation algorithms following the mechanics of natural genetics and natural selection. The broad steps in optimisation using GA are illustrated in Fig. 2.3 and works according to the following sequences:

(I) *Population Initialization:* Population size in a GA is determined by number of strings in a population and each string contains the same number of substrings as that of the constraining variables. Initial population for GA is generated randomly within the range of the constraints while the population size is determined by user.

(II) *Objective Function Evaluation:* Values of substrings in every string of the initial population are further used to evaluate the corresponding D_i's and combined objective function value $\mathbf{J}$ using Eq. (2.24).

(III) *Fitness Scaling:* It converts the objective function values obtained in a particular generation to the scaled values within a range, known as fitness function values (f_i). Presently, *Rank Scaling function* is used for the purpose. The advantage of rank scaling is that the objective function does not need to be accurate as long as it can provide correct ranking information (Jang, Sun, Mizutani 2007).

(IV) *Selection or Reproduction:* Selection operator (Deb 2005) identifies good (above average) solutions in a population, generate multiple copies of good solutions and eliminate bad solutions from population proportionately

Fig. 2.3 Flow Diagram of GA

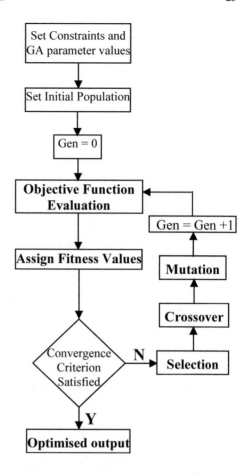

keeping population size constant. If the average fitness of all population members is f_{avg}, a solution with a fitness f_i gets an expected f_i/f_{avg} number of copies in the mating pool. In the present work, *Roulette wheel selection operator* is used for developing the mating pool through selection of good strings in a population.

(V) ***Crossover:*** In crossover operator, new strings are created by exchanging information among the strings of the mating pool. The *two-point crossover operator* with a user defined crossover fraction value is used in present model. It randomly picks two parent strings from the mating pool and swaps the part of strings between two randomly selected crossover points to generate two new strings or children.

(VI) ***Mutation:*** *Uniform mutation operator* is further applied on the modified population after reproduction and crossover to generate new strings known as mutation children by small random changes in the individuals within the population through a user defined mutation rate.

Table 2.1 GA operators
with set values/options

Sl. No.	GA operators	Set values/options
1	Size of populations	Variable
2	Number of stall generations	Variable
3	Fitness function	Rank scaling
4	Selection function	Roulette wheel
5	Crossover function	Two point
6	Crossover fraction	Variable
7	Mutation function	Uniform
8	Mutation rate	Variable

(VII)   ***Termination of computation with optimized output:*** Potential of the new population generated by the GA operators is evaluated again through generation of output variables in step (*II*) and fitness values are calculated in step (*III*). If best fitness value calculated in step (*III*) is improved, GA enters into next generation (iteration) to compute GA operators with new population. The computation stops if there is no further improvement in the best fitness value for a certain number of consecutive iteration (or reproduction of new generations) cycles. This termination number of iterations is called 'stall generations'. At stall generation, value of combined objective function **J** is considered as optimum value of response. The corresponding operational input parameters (X_i's) which yield this optimal response (or values of output parameters, D_i's) are identified and will form the optimum input parameter set for the process at hand.

During GA optimisation in the present model, size of populations, number of stall generations, crossover fraction and mutation rate are considered as variables and may be varied during optimisation of parameters of different manufacturing processes. Options for other GA operators are kept fixed and given in Table 2.1.

2.6   Optimisation Using Simulated Annealing

Simulated Annealing (SA) is a random search technique that resembles the cooling process of the molten metal through annealing and this corresponds to attainment of equilibrium state with minimum possible free energy associated with the crystallised state upon solidification. Achieving the equilibrium state by a very slow cooling rate is known as annealing in metallurgical practice. SA simulates this process of slow cooling to achieve minimum of a function value in minimisation problem. The cooling phenomenon is simulated by controlling a temperature-like parameter introduced with the concept of the Boltzmann probability distribution already mentioned in Sect. 1.3.3. SA uses this concept of minimum energy by representing energy **J** with the function value to be optimised in dependence of parameters X_i and with a control parameter likened to the temperature T and is given by $P(\mathbf{J}) = \exp(-\mathbf{J}/T)$ only. $\mathbf{J}$ is calculated from initial or subsequent values of $\mathbf{N}$ during the iteration process. A high temperature is initially selected which ensures any arbitrary $\mathbf{N}^{[0]}$ that generates a relatively near optimised (similar to thermal equilibrium) values to start with. Neighbouring points $\mathbf{N}^{[1]}$ at the same T is then searched for further possible optimisation of the function.

The algorithm employs a random search which not only accepts changes that decrease the objective function $\mathbf{J}$ (for a minimisation problem), but also some changes that increase it. The latter are accepted with a probability of $P(\mathbf{J} + \Delta\mathbf{J}) = \min\left[1, \exp(-\Delta\mathbf{J}/T)\right]$, where $\Delta\mathbf{J}$ is the increase in $\mathbf{J}$ and T is a control parameter, which, by analogy to the original application, is known as system *'temperature'* irrespective of objective function involved. The second point $\mathbf{N}^{[1]}$ is created randomly by normal distribution and the differences in function values at these points are calculated. If the second point has a smaller value the point is accepted. Otherwise also, the point is also accepted if probability $\exp(-\Delta\mathbf{J}/T)$ is less than a random number generated between 0 and 1. If none of these conditions are satisfied, a new $\mathbf{N}^{[2]}$ is created and the search continues for next iteration value of $\mathbf{N}$ and so on till convergence criteria are fulfilled thus yielding the converged $\mathbf{N}$ as a solution point for a particular T. This acts as the initial guess $\mathbf{N}^{[1]}$ for the next temperature cycle. Simulation of Annealing process continues by changing the temperature (T) according to a chosen cooling rate (k) and T is updated during iterations using the relation $T^{t+1} = k \times T^t$. The algorithm terminates when a sufficiently small temperature is obtained or a small enough change in the function value ($\mathbf{J}$) is obtained (Deb 2005).

Working of SA is shown in Fig. 2.4 and proceeds according to the following steps:

STEP I: Optimisation begins with initial point $\mathbf{N}^{[0]}$ for $\mathbf{N}(X_1, X_2, \ldots\ldots, X_m)$. Initial temperature T for SA optimisation is determined by user and in the present study T is considered as 300 °C. To maintain slow cooling, cooling rate k is set at 0.95.

STEP II: $\mathbf{N}^{[0]}$ is then used as input to compute the output parameters D_i's, which are used to find the combined objective function $\mathbf{J}^{[0]}$ according to Eq. (2.24).

STEP III: The neighbouring new point $\mathbf{N}^{[1]}$ is determined by generating a random number according to normal distribution. It is calculated as,

$$\mathbf{N}^{t+1} = \mathbf{N}^t + \sigma \times r \tag{2.25}$$

where, r is a random number generator based on normal distribution and $\sigma = $ (upper range of constraints $-$ lower range of constraints)$/6$

Objective function value $\mathbf{J}^{[1]}$ corresponding to $\mathbf{N}^{[1]}$ is evaluated in the same way as $\mathbf{J}^{[0]}$ is calculated in *STEP II*.

STEP IV: Difference in objective function or energy values is calculated as, $\Delta\mathbf{J} = \mathbf{J}^{t+1} - \mathbf{J}^t$.

Fig. 2.4 Flow Diagram of
SA

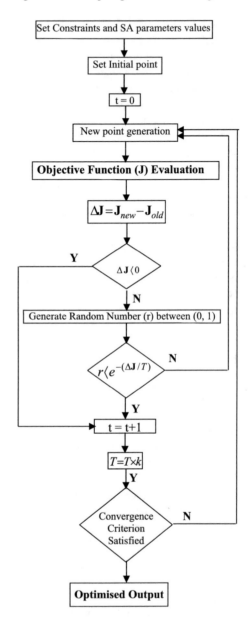

Table 2.2 SA operators with set values/options

Sl. No.	SA operators	Set values/options
1	Initial temperature (T)	300 °C
2	Cooling rate (k)	0.95
3	Number of stall iteration	variable

Acceptability of the new point is assessed as follows:

- If $\Delta J < 0$, the new point $N^{[1]}$ is accepted and program control goes to Step V.
- Else if $\Delta J \geq 0$ a random number, r is created between 0 and 1.
- It is checked whether $r < \exp(-\Delta J/T)$ is satisfied.
- If satisfied then program control goes to Step V.
- Else program control return to step III and another new point is generated.
- **STEP V:** Temperature is updated for next iteration as $T^{t+1} = k \times T^t$, where k is cooling rate.
- **STEP VI:** The computation stops if there is no further improvement in the best function value for a certain number of consecutive iteration cycles. The best function value at any iteration is the function value evaluated from the best N found so far. This termination number of iterations is called 'stall iterations', when minimum value of objective function (J), corresponding output parameters (D_i's) with optimum operational input parameters (X_i's) are determined. If termination criterion is not satisfied, program control again returns to *STEP III* and new iteration starts with new point **N** and the reduced temperature as computed in *STEP V*.

During SA optimisation in the present model, Initial Temperature (T) and Cooling Rate (k) are considered as fixed parameters. Number of stall iterations is varied for different application domains based on complexity involved. SA operators with set values are given in Table 2.2.

2.7 Integrated Methodology for Prediction and Optimisation

The foregoing discussion was based on working of different training algorithms of ANN and different optimisation methods like GA and SA. The primary objective of the present model is to integrate ANN with GA and SA optimisation techniques to develop integrated soft computing based models for prediction and optimisation of quality and mechanical properties related to LASOX cutting processes. In these integrated models, ANN is used for prediction of objective function. In each case, a single program developed in MATLAB 7.0 environment is employed for performing the prediction and optimization.

2.7.1 Modification of Objective Function

The combined objective function developed in Eq. (2.24) is used in the integrated optimisation models. As during ANN training, the input parameters are normalised within a range of 0 and 1, the range of constraining variables becomes 0-1 for the purpose of optimisation. After incorporating these modifications the optimisation problem is formulated as,

Minimise:

$$\mathbf{J} = \sum_{i=1}^{p} D_i \times W_i$$

Subject to the constraints:

$$0 \leq X_1, X_2, \ldots \ldots \ldots \ldots \ldots \ldots, X_m \leq 1 \qquad (2.26)$$

2.7.2 Integrated ANN-GA and ANN-SA Methodology for Prediction and Optimisation

The integrated ANN-GA and ANN-SA models devised here to train, predict and optimise operational parameters of LASOX cutting operations, as mentioned, by running a single program in MATLAB 7.0 environment. Program works in two steps as

(i) Determination of best ANN network configuration based on relative performances upon repeated training and testing of different architectures using different training algorithms

(ii) Optimisation with a specific optimisation algorithm (GA/SA) where the best ANN is used to compute objective functions.

A separate data file is used for storing training and testing data of ANN. The program employs the training and testing data after normalisation and calls a user defined ANN training algorithm (with user defined no. of hidden layer neurons for BPNN/spread factor values for RBFN) to perform the task of training and subsequent testing to assess the function approximation and prediction capability of a particular ANN architecture. The training and testing process is repeated for a number of architectures using different ANN training algorithms. Based on the performance during training and testing sessions, the optimum number of hidden layer neurons is determined for each ANN algorithm and forms the final architecture for the particular ANN algorithm. The architecture and the algorithm with maximum prediction accuracy are selected as best ANN for further use. The performance assessment of ANN algorithms are computed as explained in Sect. 2.2. Next, the program calls the subroutine of a particular optimisation program

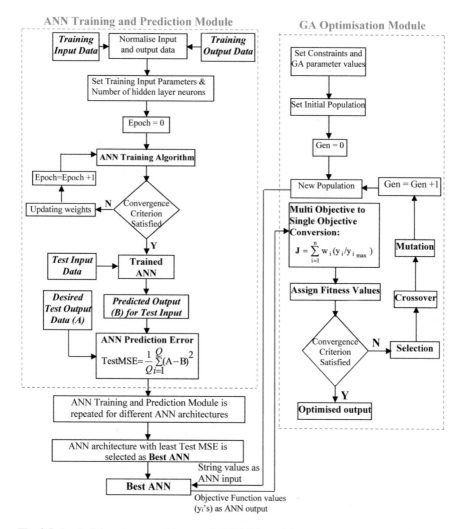

Fig. 2.5 Logical flow diagram of integrated ANN-GA model

(GA/SA) as opted by the user and starts iteration with the initial settings like initial population for GA and with initial point, $N^{[0]}$ for SA. With these initial settings, program control switches from the particular subroutine to ANN module in the main program and employs the best ANN determined earlier to predict the value of output parameters (D_i's) corresponding to initial input values. The program control then switches back to the optimisation subroutine and compute weighted sum of the ANN generated values of output parameters (D_i's) to determine combined objective function value ($\mathbf{J}$) using Eq. (2.26). The optimisation subroutine then performs the subsequent operations to improve upon the initial input values following the particular optimisation to complete a single iteration.

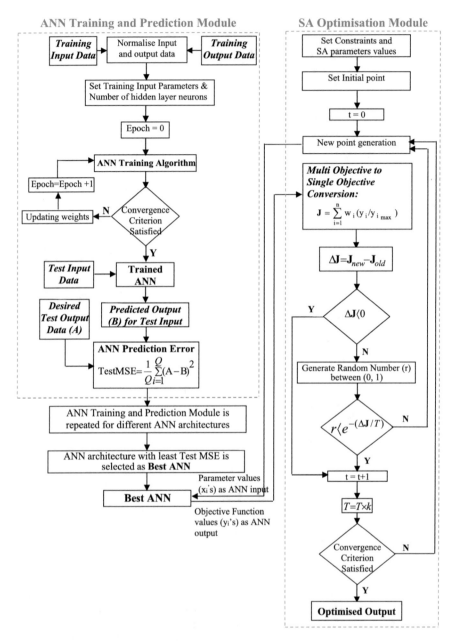

Fig. 2.6 Logical flow diagram of integrated ANN-SA model

In <u>ANN-GA integrated model</u>, GA subroutine performs the subsequent operations like selection, crossover and mutation for optimisation using the methodology explained in Sect. 2.5.

In <u>ANN-SA integrated model</u>, SA subroutine starts optimisation with $\mathbf{N}^{[0]} = [0 \quad \ldots \ldots \ldots \quad 0]$ for performing optimisation with normalised value of experimental dataset. It therefore generates the updated input point $\mathbf{N}^{[1]}$ and continues the optimisation cycle using Step II to Step IV of Sect. 2.6.

The cycle continues up to the point of appropriate convergence. On convergence the program returns the normalised optimum value $\mathbf{J}$ and corresponding value of output parameters (D_i's) with optimum operational input parameters ($\mathbf{X}_i$'s). They are further post- processed to get de-normalised value of the physical variables concerned. Logical flow diagrams for the present methods are given in Figs. 2.5 and 2.6.

2.8 Conclusion

The methodology detailed in this chapter including the development of integrated models has been implemented in the chap. 3 for prediction and optimisation of process parameters for LASOX cutting process along with comparative study of the different integrated methods. Therefore, the models and the important equations explained in this chapter have been directly referred to in the subsequent discussions for comparative performance analyses of different models.

References

Deb K (2005) Optimisation for engineering design: algorithms and examples, 8th edn. Prentice-Hall of India Private Limited, India

Hagan MT, Menhaj MB (1994) Training feed forward networks with the Marquardt algorithm. IEEE Trans Neural Networks 5(6):989–993

Haykin S (2006) Neural networks: a comprehensive foundation, 2nd edn. Pearson Education Inc, India

Hornik K, Tinchcombe M, White H (1989) Multilayer feed forward networks are universal approximators. IEEE Trans Neural Networks 2:359–366

Jang J-SR, Sun C-T, Mizutani E (2007) Neuro-fuzzy and soft computing: a computational approach to learning and machine intelligence. Pearson Education Inc., India

Kumar S (2004) Neural networks: a classroom approach, 1st edn. Tata McGraw-Hill Company Limited, pp. 304–310

Chapter 3
Modelling and Optimisation of LASOX Cutting of Mild Steel: A Case Study

Abstract Integrated ANN-GA and ANN-SA methodology are two integrated soft computing based models that can predict and optimise input-output parameters of any manufacturing process with required optimisation accuracy eliminating the need of any closed form objective functions. In the present chapter, a case study on modelling and optimisation of CO_2 LASOX cutting of mild steel plates have been carried out using the integrated ANN-GA and ANN-SA methodology to investigate efficacy of those integrated methods. In the case study cutting speed, gas pressure, laser power and standoff distance have been considered as process variables for modelling and optimisation of HAZ width, kerf width and surface roughness. 36 different ANNs have been trained and tested for ANN modelling. Finally, 4-8-3 network during training and testing through BPNN with BR results best prediction performance with MSE of 8.63E−04. Prediction capability of the best ANN (4-8-3) is found superior compared to the second order regression models developed for the purpose and is used in integrated models for optimisations. During optimisation, ANN-SA is found to show best optimisation performance with maximum absolute % error of 5.18% during experimental validation. Optimum cut quality is produced by low gas pressure and high cutting speed, laser power and stand off distance.

Keywords CO_2 LASOX cutting · Modelling · Optimisation · Artificial neural networks (ANN) · Genetic algorithms (GA) · Simulated annealing (SA) Integrated ANN-GA methodology · Integrated ANN-SA methodology

3.1 Introduction

In laser cutting of mild and carbon steels, oxygen is widely used in industries as assist gas where the release of exothermic energy from the combustion of iron contributes excess energy to the process, enhancing process speed and cut depths in ferrous metals. But the process behaves well during cutting of steel sections up to 15 mm thickness with sub 2 kW lasers. But cutting of steel sections having thickness above 15 mm require laser power substantially above 2 kW with reduction in process stability. In recent past, O'Neill and Gabzdyl (2000) has developed laser assisted oxygen cutting (LASOX) method for thick section cutting of mild steel with low power laser. In LASOX cutting, a short focal length lens in nozzle assembly results in a diverging beam on the surface of the steel with the co-axial oxygen stream. In this method laser beam interaction area should be greater than gas jet interaction area which is known as LASOX condition. As a result, 80% of the oxidation energy is used for cutting and a low power laser (sub 2 kW) should be sufficient to cut a steel plate up to 40 mm thickness. Detailed working of the LASOX cutting has been explained in Chap. 1. Controllable parameters of LASOX cutting involves laser power, cutting speed, assist gas pressure, stand off distance, thickness of work material etc and output parameters are kerf width, kerf deviation, heat affected zone, material removal rate, other metallurgical properties etc. However, those input parameters bear complex relationship among them and soft computing techniques can be employed for modelling input-output relationship and optimisation of such complex process. The model developed can act as an efficient supervisory control for online process monitoring to maintain the product quality in optimised region. Soft computing techniques like, Artificial Neural Networks (ANN), Genetic Algorithm (GA) and Simulated Annealing (SA) can be used for that purpose. Moreover, integration of such techniques eliminate the need of closed form objective function for optimisation and complete modelling of input-output relationship with subsequent optimisation by running a single program. That methodology has been discussed in detail in Chap. 2 with detailed formulation of integrated models like ANN-GA and ANN-SA respectively.

In the present chapter, an experimental dataset on LASOX cutting of mild steel is employed for modelling and optimisation of the process and kerf width, HAZ width and surface roughness are measured as process output parameters during the study. Input-output dataset generated from experiment is analysed in two steps:

(I) Selection of best ANN model and architecture based on prediction capability on training and testing of different architectures of ANN models (BPNN with LM, BPNN with BR and RBFN)

(II) Determination of best integrated optimisation model (ANN-GA and ANN-SA) for optimisation of controllable cutting parameters where best ANN model has been employed for determination of combined objective function value.

Basic details of experiment, training and prediction with ANN, comparison of best ANN with regression model and optimisation using integrated optimisation models have been discussed in the subsequent sections.

3.2 Present Problem and Dataset for Modelling

The experimental data for the present work has been obtained from the experiments carried out by Sundar et al. (2009) on LASOX cutting operation of 40 mm thick mild steel plate using an indigenous CO_2 laser machine at School of Laser Science and Engineering, Jadavpur University, India. The independent input process parameters for the experiment conducted are cutting speed (V, mm/min), gas pressure (PR, bar), laser power (P, W) and stand off distance (F, mm). Range of the process parameters in the experiment had been selected through a series of trial runs. The numerical values of control factors at different levels for mild steel sheets are shown in Table 3.1. Experiments have been conducted in random order to avoid any systematic error. Resulting dataset contains 25 non-replicated experimental data.

The heat affected zone (HAZ) is the area of base material near the surface of cut where microstructure and properties are altered by heat produced during cutting/welding. A good cutting operation demands a smaller HAZ width, a narrow kerf width and low surface roughness for the cut surface. Therefore, in the present study performance of LASOX cutting operation is evaluated by measurement of the output variables namely, (i) HAZ (heat affected zone) width (H, mm), (ii) kerf width (K, mm) and (iii) surface roughness (R, μm). Best cutting quality is indicated by achieving the minimum values for all the three output variables of kerf width, surface roughness and HAZ width. This would mean a multi objective optimisation in respect of input operating parameters. In the present study, however, this multi-objective problem has been reduced to a single objective one which is created by a linear combination of the three-output parameter with suitable weights attached to them, as discussed later. The root mean square value of surface roughness (R) is measured by a portable surface roughness tester (Taylor–Hobson Surtronic 3+ surface profilometer). Surface roughness measurements have been taken along the length of cut at three places across the depth and average of these three values are recorded. Average HAZ widths of top and bottom surfaces are measured by optical microscope (least count 0.001 mm) and the average kerf width of top and bottom surfaces is measured by low resolution microscope fitted with x-y table. The 25 nos. of experimental datasets are used in the present study for working with the GA-ANN and SA-ANN hybrid model. Out of these 25 datasets, 20 sets have been considered for ANN training and the rest are used for ANN testing. Training and

Table 3.1 Levels of experimental input parameters

Experimental input parameters	Levels				
	1	2	3	4	5
Cutting speed V (mm/min)	400	410	420	430	440
Gas pressure PR (bar)	5.5	6	6.5	7	7.5
Laser power P (W)	900	950	1000	1050	1100
Stand off F (mm)	2	2.2	2.4	2.6	2.8

testing datasets are furnished in Table 3.2. Thus, input vector (**X**) and output vector
(**D**) for the present problem is given by,

$$\mathbf{X} = \begin{bmatrix} V & PR & P & F \end{bmatrix}, \quad \mathbf{D} = \begin{bmatrix} H & K & R \end{bmatrix} \tag{3.1}$$

Table 3.2 Experimental dataset for ANN training and testing

Training data							
Exp. no.	Input parameters				Output parameters		
	Cutting speed V (mm/min)	Gas pressure PR (bar)	Laser power P (W)	Stand off distance F (mm)	HAZ width H (mm)	Kerf width K (mm)	Surface roughness R (μm)
1	410	6	950	2.2	2.53	4.505	31.5
2	410	7	950	2.2	2.655	4.955	30.56
3	430	7	950	2.2	2.735	4.895	29.56
4	430	6	1050	2.2	2.52	4.03	28.3
5	410	7	1050	2.2	2.465	5.055	26.5
6	410	6	950	2.6	2.225	4.18	32.1
7	430	6	950	2.6	2.505	3.975	30.75
8	430	7	950	2.6	2.66	4.49	28.5
9	410	6	1050	2.6	2.095	3.965	28.5
10	430	6	1050	2.6	2.36	3.81	27
11	410	7	1050	2.6	2.315	4.605	25.3
12	430	7	1050	2.6	2.525	4.29	23
13	400	6.5	1000	2.4	1.585	5.165	22
14	420	5.5	1000	2.4	1.675	3.975	30.85
15	420	7.5	1000	2.4	2.625	4.89	22.3
16	420	6.5	900	2.4	2.565	4.895	31.8
17	420	6.5	1100	2.4	1.94	4.015	23.2
18	420	6.5	1000	2	2.52	4.1	27.2
19	420	6.5	1000	2.8	1.865	4.54	26.1
20	420	6.5	1000	2.4	2.1	4.2	27.2
Test data							
1	410	6	1050	2.2	2.385	4.31	29.76
2	430	7	1050	2.2	2.64	4.49	24.36
3	410	7	950	2.6	2.54	4.84	31
4	440	6.5	1000	2.4	2.855	3.515	32
5	430	6	950	2.2	2.74	4.23	31.5

3.3 Selection of Model for Prediction of Cutting Quality

In the present study, different architectures of three feed forward ANN schemes, namely, BPNN with LM, BPNN with BR and RBFN have been employed for training and testing/prediction of cutting quality and material removal rate (Chaki and Ghosal 2010). Detailed algorithms of ANN models have been already explained in Sect. 2.2. Finally, based on prediction (or testing) performance, best ANN architecture is selected and compared with prediction performance of a regression model developed from experimental dataset. The selected best model is employed for subsequent analysis.

3.3.1 Application of ANN Model

In the present analysis, there are four experimental input parameters and three output parameters as indicated by input vector **X** and output vector **D** in Eq. (3.1). Therefore, present ANN model has 4 nodes in input layer and 3 nodes in output layer. Number of hidden layer neurons is considered as design parameter for BPNN and is varied to form different ANN architectures, while in case of RBFN, spread factor of Gaussian activation functions in hidden layer is considered as a design parameter and is varied to obtain different RBFN architectures. Architecture of the ANN for present problem is given in Fig. 3.1. Experimental dataset in Table 3.2 is used for training and testing. Range of experimental input and output parameters as given in Table 3.3 are used for normalisation of dataset. Parameters setting for ANN training, prediction performances of different architectures and their comparative studies are detailed in subsequent sections.

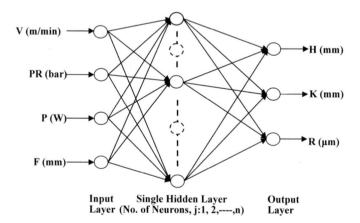

Fig. 3.1 Schematic diagram of ANN model

Table 3.3 Range of experimental input and output parameters

	Experimental input parameters				Experimental output parameters		
	Cutting speed V (mm/min)	Gas pressure PR (bar)	Laser power P (W)	Stand off distance F (mm)	HAZ width H (mm)	Kerf width K (mm)	Surface roughness R (μm)
Experimental value (before normalisation)							
Minimum value	400	5.5	900	2	1.585	3.515	22
Maximum value	440	7.5	1100	2.8	2.855	5.165	32.1
Normalised value							
Minimum value	0	0	0	0	0.555	0.681	0.685
Maximum value	1	1	1	1	1	1	1

3.3.1.1 Application of BPNN

During application of BPNN with LM algorithm and BR algorithm for training and testing, 13 different ANN architectures with number of hidden layer neurons varied from 6 to 18 have been employed. Other necessary training parameters have been detailed in Sect. 2.2.1.6. Working of BPNN has been detailed in Sect. 2.2.1. MSE is considered as performance index during training and testing and computed using Eqs. (2.5) and (2.11).

3.3.1.2 Application of RBFN

In the application of RBFN as exact interpolator for evaluation of prediction performance, spread factor is considered as design parameter of the network and is varied from 0.1 to 1.0. Other specifications such as weight initialisation, normalisation of dataset, performance indices are considered as same as those in the BPNN training. Working of RBFN in general has been detailed Sect. 2.2.2.

3.3.2 Application of Regression Model

A multivariable regression model is developed based on the experimental dataset (Table 3.2) to find out a relationship between four input process variables such as cutting speed (V, mm/min), gas pressure (PR, bar), laser power (P, W) and stand off distance (F, mm) with three output parameters of HAZ width (H, mm), (ii) kerf width (K, mm) and (iii) surface roughness (R, μm). As the training data for ANN modelling in Table 3.2 is used for the developing the regression model, it does not contain any replicated experimental run. Thus, ANOVA with Lack of Fit Test could not be conducted. Adequacy of the model has been tested by Regression coefficient (R-square) value of the regression equations being developed. Initially, a linear model has been developed but was later rejected based on very low R-square value. The model has been developed by using MINITAB 15 Software and the equations developed for H, K and R, are given below:

$$
\begin{aligned}
H = {}& 173.1038 - 0.4551 \times V + 0.5323E - 3 \times V^2 - 1.9104 \times PR \\
& + 0.2225 \times PR^2 - 0.1195 \times P + 0.3250E - 4 \times P^2 \\
& - 10.6102 \times F + 1.6563 \times F^2 - 0.0053 \times V \times PR \\
& + 0.6406E - 4 \times V \times P + 0.0022 \times PR. \times P - 0.0011 \times V \times F \\
& - 0.2943 \times PR. \times F + 0.0045 \times P \times F
\end{aligned}
\tag{3.2}
$$

$$
\begin{aligned}
K = {}& 221.4730 - 0.9987 \times V + 0.0013 \times V^2 \\
& + 3.2243 \times PR + 0.1792 \times PR^2 - 0.0237 \times P \\
& + 0.2017E - 4 \times P^2 - 2.1918 \times D + 0.4167 \times F^2 \\
& - 0.0119 \times V \times PR - 0.3888E - 4 \times V \times P \\
& - 0.4967E - 3 \times PR \times P - 0.0031 \times V \times F \\
& + 0.2185 \times PR \times F - 0.1229E - 3 \times P \times F
\end{aligned}
\tag{3.3}
$$

$$
\begin{aligned}
R = {}& -1124.9090 + 5.79262 \times V - 0.00568 \times V^2 \\
& + 17.75472 \times PR + 0.80889 \times PR^2 - 0.18425 \times P \\
& + 0.17334E - 3 \times P^2 + 2.73837 \times F + 5.52431 \times F^2 \\
& - 0.04544 \times V \times PR - 0.57062E - 3 \times V \times P - 0.00288 \times PR \times P \\
& - 0.06594 \times V \times F - 4.00869 \times PR \times F + 0.02259 \times P \times F
\end{aligned}
\tag{3.4}
$$

A near unity R-square value should indicate good accuracy of fit of the equations of a regression model. Here, R-square values for the output parameters, HAZ width (H), Kerf Width (K) and Surface Roughness (R) are found to be 0.6691, 0.8509 and 0.7936 respectively. They indicate moderate accuracy for regression model of HAZ width (H) and adequate accuracy for Kerf Width (K) and Surface Roughness (R) models. Relatively low R-square value for HAZ may be due to the inherent noise in dataset, induced during experimentation and post experimentation measurement. But, as the LASOX dataset has been taken from a published literature there were no possibility to reexamine the experimental dataset. However, as the R-square value for other factors are acceptable, further analysis has been carried out. Further, the models are tested with test data from Table 3.2 to evaluate prediction capability of the regression model.

3.3.3   Results and Discussion

3.3.3.1   Performance Analysis of ANN Training and Testing

In the present analysis, ANN training and testing has been performed with training and test data as given in Table 3.2. Performances (in terms of MSEs) of different architectures for BPNN with LM algorithm and BPNN with BR during training and testing are shown in Figs. 3.2 and 3.3. Prediction performances of RBFN architectures have been detailed in Fig. 3.4.

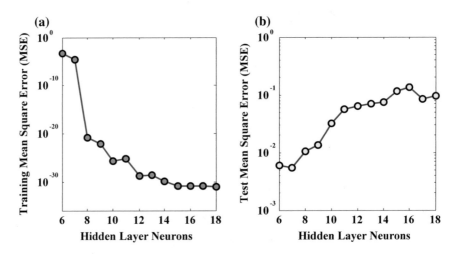

Fig. 3.2 Variation of training and testing performance with hidden layer neurons during training and testing using BPNN with LM

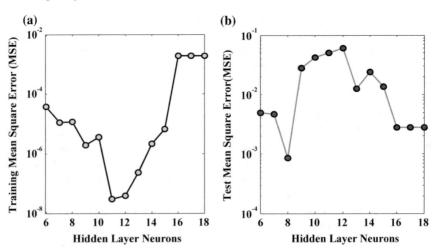

Fig. 3.3 Variation of training and testing performance with hidden layer neurons during training and testing using BPNN with BR

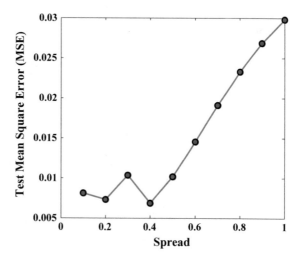

Fig. 3.4 Variation of test MSE with spread during testing using RBFN

- *Performance Analysis of BPNN with LM*

Figure 3.2a shows training performance (i.e. training MSE) improves rapidly with increase in number of hidden layer neurons from 4-8-3 networks but with a more or less gradual deterioration of prediction performance (i.e. test MSE) as obtained from Fig. 3.2b. It may be due to the over fitting of the networks. During any over-fitting in ANN, training data fits very accurately and explains the relationship between the predictor and response variables very well, but results in poor prediction capability during testing with test dataset as the correct trend may not be captured for precise interpolation or extrapolation. The possibility of over fitting is further indicated from a sudden decrease in training time with increase in hidden layer neurons (Table 3.4) which is contradictory with general trend of increase in training time with network complexity. Table 3.4 shows significant decreases in training time from 4-8-3 network indicating possible over fitting. However, prediction capability being the primary objective of a trained ANN, performance of a particular ANN during testing with test data has been considered as the yardstick for selecting the best ANN architecture. Best prediction performance is indicated by minimum testing MSE obtained by networks and given in Fig. 3.2b. Therefore, despite of showing best training MSE (9.90E−32), the 4-18-3 network results in poor prediction performance with test MSE of 9.65E−02 and not considered as best ANN. 4-7-3 network is considered as best ANN here with minimum test MSE of 5.40E−03 (Table 3.4). Therefore, the network with 7 hidden layer neurons (i.e. 4-7-3 network) is considered as best ANN with best prediction performance during training and testing using BPNN with LM algorithm.

- *Performance Analysis of BPNN with BR*

Similar training and testing operation is conducted using BPNN with BR. Details of training and testing performance is given in Table 3.4. Figure 3.3a shows

Table 3.4 Training and testing performance of different ANN architectures using BPNN with LM algorithm and BPNN with BR algorithm

Sl. no.	Network	Training algorithm					
		BPNN with LM			BPNN with BR		
		Training		Testing	Training		Testing
		MSE	Time (s)	MSE	MSE	Time (s)	MSE
1	4-6-3	5.56E−04	22.85	5.96E−03	3.87E−05	5.33	5.02E−03
2	4-7-3	2.09E−05	12.62	*5.40E−03*	1.11E−05	5.45	4.75E−03
3	4-8-3	1.75E−21	5.24	1.06E−02	1.16E−05	5.82	*8.63E−04*
4	4-9-3	6.08E−23	1.87	1.33E−02	1.90E−06	6.00	2.80E−02
5	4-10-3	1.84E−26	2.75	3.18E−02	3.55E−06	5.83	4.17E−02
6	4-11-3	5.74E−26	1.26	5.70E−02	3.00E−08	6.16	5.03E−02
7	4-12-3	1.81E−29	1.10	6.32E−02	4.00E−08	7.28	6.04E−02
8	4-13-3	2.62E−29	1.21	7.06E−02	2.30E−07	6.59	1.25E−02
9	4-14-3	1.25E−30	1.18	7.40E−02	2.10E−06	7.77	2.39E−02
10	4-15-3	1.52E−31	1.15	1.14E−01	6.52E−06	8.02	1.36E−02
11	4-16-3	1.21E−31	1.96	1.35E−01	1.87E−03	7.38	2.84E−03
12	4-17-3	1.54E−31	1.12	8.48E−02	1.87E−03	7.88	2.83E−03
13	4-18-3	9.90E−32	0.60	9.65E−02	1.87E−03	8.61	2.84E−03

continuous improvement in training performance up to training with 4-11-3 architecture. Then on increasing hidden layer neurons from 12 to 16, a gradual deterioration in training performance is observed (Fig. 3.3a). Though 4-11-3 network shows best training performance, test MSE obtained by it is 5.03E−02 which is far inferior to best test MSE of 8.63E−04 obtained by 4-8-3 network. Thus, 4-8-3 network is considered as ANN with best prediction performance during training and testing using BPNN with BR algorithm.

- *Performance Analysis of RBFN*

Finally, RBFN as exact interpolator is employed in the present problem for any possible improvement in prediction capability. Spread factor has been varied from 0.1 to 1 and the corresponding prediction performance in terms of test MSE has been given in Table 3.5. It is observed from the   Fig. 3.4 that, best prediction MSE of 6.83E−03 is obtained by the network with 0.4 spread and after that test MSE gradually increases with spread factor. Therefore, the network with 0.4 spread factor is considered as ANN with best prediction performance during application of RBFN.

- *Comparative study of ANN models*

Finally, a comparative study has been made using network with best prediction performance obtained from each of the training algorithm. Network with best prediction performance are obtained as 4-7-3 network, 4-8-3 network and network

Table 3.5 Performance of different ANN architectures during training and testing using RBFN as exact interpolator

Sl. no.	Spread	Operation time (s)	Prediction/test MSE
1	0.1	0.232221	8.08E−03
2	0.2	0.134817	7.25E−03
3	0.3	0.151865	1.03E−02
4	0.4	0.142103	*6.83E−03*
5	0.5	0.140122	1.01E−02
6	0.6	0.145935	1.45E−02
7	0.7	0.120576	1.91E−02
8	0.8	0.139172	2.33E−02
9	0.9	0.141285	2.68E−02
10	1	0.142673	2.98E−02

Table 3.6 Comparison of prediction performance of different ANN models

Training algorithm	Design parameters	Optimum value of design parameters	Best network	Prediction performance Test MSE
BPNN with LM	Hidden layer neurons	7	4-7-3	5.40E−03
BPNN with BR	Hidden layer neurons	8	4-8-3	8.63E−04
RBFN as exact interpolator	Spread of Gaussian activation function	0.4	4 input neurons, 3 output neurons and 0.4 spread	6.83E−03

with 0.4 spread during training and testing through BPNN with LM algorithm, BPNN with BR algorithm and RBFN algorithm, respectively. Table 3.6 presents a comparative study from which it is quite clear that 4-8-3 network during BPNN with BR training and testing yields best prediction performance compared to others with test MSE of 8.63E−04. Therefore, 4-8-3 network and BPNN with BR training algorithm is considered as best ANN among all 36 trained and tested network architectures and used for subsequent comparison with regression model developed.

3.3.3.2   Comparison of Best ANN Model with Regression Model

- *Prediction performance of best ANN*

From previous analysis, 4-8-3 network during training and testing with BPNN with BR is considered as best ANN. Prediction capability of that model is detailed in this

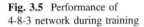

Fig. 3.5 Performance of 4-8-3 network during training

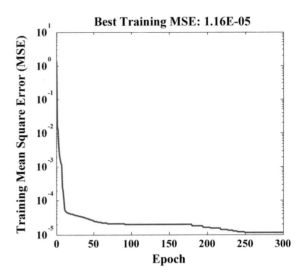

section. Training performance of the model is provided in Fig. 3.5. During testing with test data, prediction capability of the best ANN ie 4-8-3 network is assessed by calculating absolute % error and mean absolute % error in prediction for every test data after corresponding de-normalisation through Eqs. (2.13) and (2.12). During testing with test data it is observed from Table 3.7 that best ANN can reduce mean absolute % error up to 6.03, 4.33 and 5.79% corresponding to HAZ width, kerf width and surface roughness respectively. Further, overall mean absolute % error combining all output characteristics (i.e. H, K and R) is determined through Eq. (2.12) and for best ANN that value is 5.38%. Moreover, the maximum absolute % errors for them during ANN testing are 8.43, 12.8 and 9.52% respectively as furnished in Table 3.7. Though absolute % errors are somewhat higher in some of the test cases, in some cases they are reasonably low and the mean absolute error barely exceeds 6% for all the output variables. The relatively large percentage errors may be due to effect of uncontrollable noise factors in the experimental dataset.

Table 3.7 Performance of 4-8-3 networks during testing with testing input data

Exp. no	Experimental output			Predicted output of ANN model			Absolute error			Absolute % error		
	H (mm)	K (mm)	R (μm)	H (mm)	K (mm)	R (μm)	H (mm)	K (mm)	R (μm)	H (mm)	K (mm)	R (μm)
1	2.385	4.31	29.76	2.16	4.218	27.33	0.225	0.092	2.43	9.43	2.13	8.17
2	2.64	4.49	24.36	2.649	4.397	24.86	0.009	0.093	0.5	0.34	2.07	2.05
3	2.54	4.84	31	2.326	4.968	28.05	0.214	0.128	2.95	8.43	2.64	9.52
4	2.855	3.515	32	2.683	3.965	29.4	0.172	0.45	2.6	6.02	12.8	8.13
5	2.74	4.23	31.5	2.577	4.146	31.84	0.163	0.084	0.34	5.95	1.99	1.08
Mean absolute % error										6.03	4.33	5.79

Moreover, the absolute error matrix for each of the output parameters as given in Table 3.7 indicates very small measurable values (mm, μm) of errors for the output parameters. The order of magnitude of the errors indicates fairly good accuracy of ANN during prediction for a new set of input parameters.

Further a testing analysis has been conducted by feeding all experimental data to trained 4-8-3 network and predicted outputs by trained ANN is plotted against desired or experimental output as given in Fig. 3.6. The predicted outputs are plotted versus the desired output as open circles and a linear fit is drawn with solid line. The best linear fit (output equal to desired) is indicated by a dashed line. Correlation coefficient (R-value) is computed as degree of linear association between the predicted and desired outputs. An R-value close to 1 indicates good fit. In Fig. 3.6, R-value for HAZ width is near 0.9 and R-value for other two outputs are more than 0.9 which indicate good prediction capability.

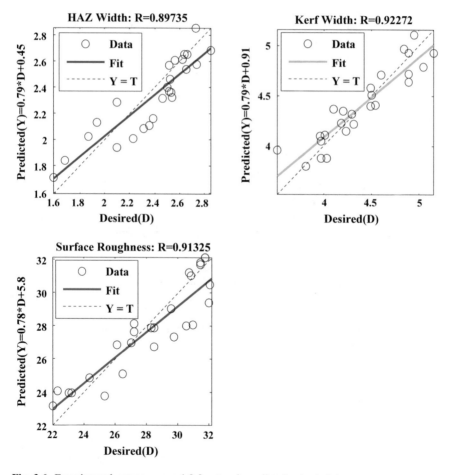

Fig. 3.6 Experimental output versus 4-8-3 network predicted output plot

- *Prediction performance of regression model*

Finally, prediction capability of the regression equations developed in Sect. 3.3.2 is
tested with same test dataset that has been used for testing of 4-8-3 network.
Predicted outputs with absolute % errors are given in Table 3.8. Absolute % error is
calculated from Eq. (2.22). Maximum absolute % errors for regression models are
30.80, 20.03 and 28.02% respectively corresponding to HAZ width, kerf width and
surface roughness. Table 3.8 also shows that, mean absolute % errors for them are
11.06, 6.55 and 11.50% respectively.

- *Comparison of prediction performance*

Table 3.8 shows that, maximum absolute % errors obtained during testing of
regression model are much higher compared to maximum absolute % errors
obtained during ANN testing. A comparative study of predicted output of best ANN
and regression models are provided in Fig. 3.7 from which it is clear that according
to the order of magnitude of errors, best ANN shows substantially better prediction
capability compared to regression models. Therefore, 4-8-3 network trained by
BPNN with BR algorithm is further employed for evaluation of objective functions
during optimisation with GA and SA algorithm in subsequent sections.

3.4   Optimisation of Cutting Quality

In the present work optimisation of process parameters is performed by employing
three different integrated soft computing-based models such as, ANN-GA and
ANN-SA. ANN module has been already detailed in Sect. 3.3 where it has been
found that 4-8-3 network trained by BPNN with BR algorithm shows best prediction capability and considered as best ANN. In the present section, that best
ANN has been employed for computation of objective function values during
iteration of GA/SA optimisation module. Working of ANN-GA and ANN-SA
models have been already detailed in Sect. 2.7. GA and SA optimisation
methodology with parameter settings have been separately explained in Sects. 2.5
and 2.6. Formulation of objective function and application of the optimisation
modules (GA and SA) have been detailed in the following sections. Computation
for optimisation modules have been carried out on a desktop Pentium IV, 3 GHz
and 512 MB PC.

3.4.1 Determination of Objective Functions

In LASOX cutting, best cut quality is indicated by narrow kerf, small heat affected
zone (HAZ) and smooth surface finish. Therefore, in the present optimisation model
best cut quality can be achieved by minimising HAZ width (H), kerf width (K) and

Table 3.8 Performance of regression model during testing with testing input data

Exp. no.	Experimental output			Predicted output of regression model			Absolute error			Absolute % error		
	H (mm)	K (mm)	R (µm)	H (mm)	K (mm)	R (µm)	H (mm)	K (mm)	R (µm)	H (mm)	K (mm)	R (µm)
1	2.385	4.31	29.76	1.823	4.205	26.79	0.562	0.105	2.97	30.80	2.49	11.10
2	2.64	4.49	24.36	2.736	4.325	23.89	0.096	0.165	0.47	3.52	3.80	1.97
3	2.54	4.84	31	2.246	5.084	27.02	0.294	0.244	3.98	13.10	4.80	14.72
4	2.855	3.515	32	2.696	4.395	24.99	0.159	0.880	7.00	5.90	20.03	28.02
5	2.74	4.23	31.5	2.686	4.300	32.04	0.054	0.070	0.54	2.00	1.64	1.68
Mean absolute % error										11.06	6.55	11.50

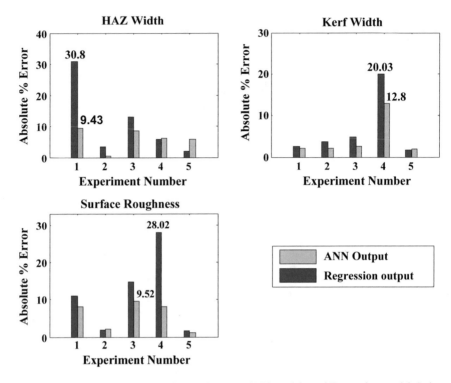

Fig. 3.7 Comparison of absolute % error between ANN models and Regression model during testing with testing input data

surface Roughness (R). The present problem is, thus, essentially multi-objective in nature and the basic objective functions can be written in generic form as,

The objective functions:

$$Minimise: \mathrm{H}(\mathrm{V,\ PR,\ P,\ F}),\ \mathrm{K}(\mathrm{V,PR,P,F}),\ \mathrm{R}(\mathrm{V,PR,P,F})$$

Subject to the *constraints*:

$$\mathrm{V_{min}} \leq \mathrm{V} \leq \mathrm{V_{max}}, \mathrm{PR_{min}} \leq \mathrm{PR} \leq \mathrm{PR_{max}}, \mathrm{P_{min}} \leq \mathrm{P} \leq \mathrm{P_{max}}, \mathrm{F_{min}} \leq \mathrm{F} \leq \mathrm{F_{max}} \quad (3.5)$$

Where, V, PR, P, F are the LASOX cutting input parameters and H, K and R represent the cutting quality parameters as output.

ANN training module normalises the input and output dataset between 0 and 1 for better training performance. The trained ANN can compute and predict output only within the range of 0 and 1. As such, the input range of constraining variables has been normalised between 0 and 1 (Table 3.3).

The present study reduces the multi-objective problem to a single objective optimisation problem by using a function **J** that represents cutting quality factor as a weighted sum of the output parameters. Therefore, the objective function detailed in Eq. (3.5) is modified for optimisation as follows:

The objective functions:

$$Minimize: \mathbf{J} = W_1 \times (H/H_{max}) + W_2 \times (K/K_{max}) + W_3 \times (R/R_{max})$$

Subject to the *constraints*:

$$0 \leq V, PR, P, F \leq 1 \qquad\qquad (3.6)$$

H_{max}, K_{max} and R_{max} values can be obtained from Table 3.3. W_1, W_2 and W_3 represent the parameters to denote weight of each output. '+'ve sign before an output parameter indicates that the parameter is to be minimised. Here, choice of the weights is completely subjective and depends on the requirement of user. Equal weight of 0.33 is considered for all output variables H, K and R in the present problem.

3.4.2   Application of Optimisation Modules

In *GA optimisation module*, every string in a population contain four substrings that indicate the constraining variables of V, PR, P and F. Options for basic GA operators as provided in Table 2.1 have been employed here for optimisation.

The other variable operator values for the present optimisation are given below:

Population size: 200
Number of Stall generations: 20
Crossover Fraction: 0.5
Mutation Rate: 0.02
Optimisation stops when stall generation criterion is reached.

In *SA optimisation module*, optimisation starts with the initial point **N** (V, PR, P, F) and terminates upon satisfying the stall iteration criterion. Number of 'stall iterations' in present problem is fixed at 50. Values of other operators are provided in Table 2.2. Optimisation stops when stall generation criterion is reached.

On convergence, optimisation modules (GA and SA) returns combined objective function **J**, H, K, R and corresponding optimum operational parameters of LASOX cutting i.e. V, PR, P and F.

3.4.3   Results and Discussion

3.4.3.1   Performance Analysis of ANN-GA Optimisation

In integrated ANN-GA optimisation, search ends after 54 generations (iterations) when 'stall generation' criterion is reached. Figure 3.8a shows the plot of best function value in each generation versus generation number as well as convergence nature of the problem. During minimisation of cutting quality factor (**J**) the optimised value of **J** is found as 0.7948. The optimised values of H, K and R, after de-normalisation are found to be 2.210 mm, 3.939 mm and 28.0 μm with corresponding operational input parameter (V, PR, P and F) values as, 428.44 mm/min, 6.0654 bar, 1047.8 W and 2.5636 mm respectively. These operational input parameter values closely match with a specific input parameter setting of available experimental data (Table 3.2) i.e. 430 mm/min, 6.0 bar, 1050 W and 2.6 mm and corresponding experimental outputs are 2.36 mm, 3.81 mm and 27.0 μm. The accuracy of ANN-GA model is evaluated by calculating the absolute % error between outputs predicted by ANN-GA model and experimental data and furnished in Table 3.9. It is observed that, the ANN-GA model can predict optimised output with absolute % error of 6.36, 3.39 and 3.65% for H, K and R. Total computational time of ANN-GA model combining training and prediction with best ANN and optimisation with GA is found to be 152.11s only.

3.4.3.2 Performance Analysis of ANN-SA Optimisation

Optimisation using present ANN-SA model is ended after 121 iterations when 'stall iteration' criterion is reached. Figure 3.8b shows the plot of best function value over iterations versus iteration number as well as convergence nature of the problem. During minimisation of cutting quality factor (**J**) the optimised value of **J** is found as 0.7803. During optimisation temperature T has been decreased with a very slow rate (k = 0.95) at the beginning of iterations as a step of SA algorithm. Temperature characteristic curve with respect to optimisation iterations is shown in Fig. 3.8c and it indicates that at the end of search temperature T is reduced to 0.57465 °C, which is very close to zero. Though the stall iteration is considered as a convergence criterion in the present SA optimisation process, the near zero value of T at the point of convergence indicates that, it also simultaneously satisfies the minimum temperature criterion.

Upon de-normalisation of the normalised output generated by SA, the optimised values of HAZ width (H), kerf width (K) and surface roughness (R) are found to be 2.238 mm, 3.682 mm and 27.9 μm with corresponding operational input parameter (V, PR, P and F) values as, 428.552 mm/min, 6.064 bar, 1048.6 W and 2.5714 mm respectively. The summary of the results of the optimisation is given in Table 3.10.

These operational input parameter values closely match with a specific input parameter setting of available experimental dataset (Table 3.2) i.e. 430 mm/min, 6.0 bar, 1050 W and 2.6 mm and corresponding experimental outputs are 2.36 mm, 3.81 mm and 27.0 μm respectively. The accuracy of ANN-SA model is evaluated by calculating the absolute % error between outputs predicted by ANN-SA model and experimental data and furnished in Table 3.10. It is observed that, the ANN-SA model can predict optimised output with absolute % error of 5.18, 3.37 and 3.18% for H, K and R. Total computational time of the SA-ANN model combining training and prediction with best ANN and optimisation with SA is found to be 22.82s only.

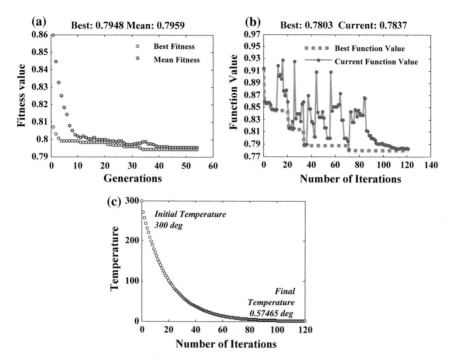

Fig. 3.8 Performance plots for GA and SA. **a** Performance of GA during optimisation. **b** Performance of SA during optimisation. **c** Change in temperature during SA optimisation

Table 3.9 Results of ANN-GA optimisation and comparison with experimental output

	Operational input parameters				Output parameters			
	V (mm/min)	PR (bar)	P (W)	F (mm)	H (mm)	K (mm)	R (μm)	Combined objective function (J)
Normalised value	0.711	0.2827	0.739	0.7045	0.77406	0.7627	0.8718	0.7948
Denormalised value	428.44	6.0654	1047.8	2.5636	2.210	3.939	28.0	
Experimental value	430	6	1050	2.6	2.36	3.81	27	
Absolute % error					6.36	3.39	3.65	

Table 3.10 Results of ANN-SA optimisation and comparison with experimental output

	Operational input parameters				Output parameters			
	V (mm/ min)	PR (bar)	P (W)	F (mm)	H (mm)	K (mm)	R (μm)	Combined objective function (J)
Normalised value	0.7138	0.282	0.743	0.7143	0.78378	0.7128	0.8679	0.7803
Denormalised value	428.552	6.064	1048.6	2.5714	2.238	3.682	27.9	
Experimental value	430	6	1050	2.6	2.36	3.81	27	
Absolute % error					5.18	3.37	3.18	

3.4.3.3 Comparison of Performance of Different Optimisation Models

Optimisation performance of ANN-GA and ANN-SA models are detailed in Sects. 3.4.3.1 and 3.4.3.2. Purpose of optimisation was to minimise cutting quality factor J and the model that produces least value of J is considered as best optimisation model. A comparative study of optimisation performance of different models has been provided in Table 3.11. It is observed from Table 3.11 that ANN-SA model produces minimum value of J (0.7803) during optimisation, compared to other optimisation models. Moreover, during comparison of ANN-SA optimised output with experimental data, absolute % error produced is fairly low (5.18, 3.37 and 3.18% respectively). It may be noted that, as the present analysis is carried out using the experimental data obtained by Sundar et al. (2009), there was no scope of conducting a validation experiment with actual optimised operational parameters obtained by ANN-SA (V = 428.552 mm/min, PR = 6.064 bar, P = 1048.6 W and F = 2.5714 mm) as shown in Table 3.10. Therefore, instead of conducting validation experiment with V = 428.552 mm/min, PR = 6.064 bar, P = 1048.6 W and F = 2.5714 mm, in the present analysis, accuracy of the model is checked with experimental output conducted with a nearby available experimental input data (V = 430 mm/min, PR = 6 bar, P = 1050 W and F = 2.6 mm). If the experiment could have been conducted with the ANN-SA optimised value of operational parameters, the experimental output might have become closer to

Table 3.11 Comparative study of optimisation performance

	Optimised output parameters			
	Absolute % error			Combined objective function (J)
	H (mm)	K (mm)	R (μm)	
ANN-GA optimisation	6.36	3.39	3.65	0.7948
ANN-SA optimisation	5.18	3.37	3.18	0.7803

optimised output with, absolute % error, further reduced. Total operational time of ANN-SA model (22.82s) is also reasonably low.

3.4.3.4 Significance of Optimised Parameters

On comparing ANN-SA optimised parameter setting with available experimental levels (Table 3.1) it can be observed that minimum value of cutting quality factor **J** can be obtained on operating with high cutting speed (V), low gas pressure (PR), high laser power (P) and high stand off distance (F) with respect of the given range in Table 3.3. However, it is observed in literature (O'Neill and Gabzdyl 2000) that, cutting quality parameters of LASOX cutting process is greatly affected by the amount of side burning of the material. During side burning the material is heated up to ignition temperature due to exothermic reaction between iron and oxygen, over a relatively wide zone around laser beam and results in high HAZ width, large kerf width and high surface roughness. The present optimisation results indicate high cutting speed (V) and low gas pressure (P). This is commensurate with the experimental finding because higher cutting speed reduces the time available for side burning and low gas pressure reduces amount of oxygen available in cutting zone for excess heat generation through exothermic reaction. As a result, side burning reduces to a great extent resulting in better cutting quality i.e. low HAZ width, low kerf width and low surface roughness. Optimisation results also indicate a high value of laser power (1050 W) which is needed for establishment and thermal stability of LASOX condition. That in turn produces better cutting quality. It is also supported by the work of O'Neill and Gabzdyl (2000). Generally, high stand off distance increases the size of beam impinging on the material with a consequent decrease in laser power per unit area, resulting high kerf width. But it is reported in work of Sundar et al. (2009) that, above a value of 2.5 mm, the effect of stand off distance is negligible. In the present optimisation results, stand off distance is 2.6 mm and thus has little effect on kerf width thereby improving the cutting quality. Moreover, increase in stand off distance causes decrease in HAZ width and surface roughness. However, effect of laser power and stand off distance are negligible compared to cutting speed and gas pressure (Sundar et al. 2009). Therefore, from above discussion it is clear that, the optimised cutting parameters are rightly predicted by the hybrid model and if applied in LASOX cutting process should produce best cutting quality.

3.5 Conclusions

In the present chapter, modelling and optimisation of cutting quality factor combining heat affected zone, kerf width and surface roughness has been carried out using integrated ANN-GA and ANN-SA model for LASOX cutting of thick mild steel plates. Different architectures of three ANN training algorithms such as BPNN with LM, BPNN with BR and RBFN have been trained and tested using experimental dataset for selection of best ANN model.

Following conclusions can be drawn on the basis of results obtained:

(i) 4-8-3 network during BPNN with BR training and testing results best prediction performance with MSE of 8.63E−04 among all 36 tested network architectures with three different training algorithms and is considered as best ANN.

(ii) 4-8-3 network has been found to reduce mean absolute % error up to 6.03, 4.33 and 5.79% corresponding to HAZ width (H), kerf width (K) and surface roughness (R) respectively.

(iii) Prediction performance of the best ANN is superior compared to a second order regression model developed.

(iv) ANN-SA is found to show best optimisation performance during optimisation with two different models like ANN-GA and ANN-SA.

(v) Best optimisation performance of combined weighted objective functions with ANN-SA can be determined if experiments are conducted with cutting speed (V) = 428.552 mm/min, gas pressure (PR) = 6.064 bar, laser power (P) = 1048.6 W and stand off distance (F) = 2.5714 mm.

(vi) On comparison of ANN-SA optimised output with experimental data, absolute % error produced is fairly low (5.18, 3.37 and 3.18% for H, K and R respectively). Total operational time is also low (22.82s). Therefore, present ANN-SA model can be used for prediction and optimisation of operational parameters of any experimental dataset with reasonable accuracy.

References

Chaki S, Ghosal S (2010) Prediction of cutting quality in LASOX cutting of mild steel using ANN and regression model. In: Proceedings of National Conference on Recent Advances in Manufacturing Technology and Management (RAMTM2010), Jadavpur University, Kolkata, 19–20 Feb, pp 163–168

Neill WO, Gabzdyl JT (2000) New developments in oxygen-assisted laser cutting. J Opt Lasers Eng 34(4–6):355–367

Sundar M, Nath AK, Bandyopadhyay DK, Chaudhuri SP, Dey PK, Misra D (2009) Effect of process parameters on the cutting quality in LASOX cutting of mild steel. Int J Adv Manuf Technol 40(9–10):865–874

Printed in the United States
By Bookmasters